FIRE AND BUILDING

A GUIDE FOR THE DESIGN TEAM

FIRE AND BUILDING

A GUIDE FOR
THE DESIGN TEAM

THE AQUA GROUP

Pictures by Brian Bagnall

GRANADA
London Toronto Sydney New York

Granada Technical Books
Granada Publishing Ltd
8 Grafton Street, London W1X 3LA

First published in Great Britain by
Granada Publishing, 1984

British Library Cataloguing in Publication Data
Fire and building.
  1. Fire prevention—Great Britain
  I. Aqua Group
  628.9'22   TH9537

ISBN 0-246-11878-4

Typeset by Columns of Reading
Printed and bound in Great Britain by Mackays of Chatham, Kent

CONTENTS

– Sprinkler systems – Foam injection – Gas and
powder systems – Smoke control – Wet and dry
risers – Hand held extinguishers and other
equipment

CONTENTS

Professional fees – Average – Excess – Long term
liability – Fire during construction – Value Added
Tax – Repair

ACKNOWLEDGEMENTS

The authors and publishers would like to thank the following:

(1) The Controller of Her Majesty's Stationery Office for permission to reproduce the *Application for a Fire Certificate* shown in Appendix 6;

(2) The GLC Department of Architecture and Civic Design for permission to reproduce the *Fire Certificate* shown in Appendix 7;

(3) Mr H.L. Malhotra for permission to use the illustration in Fig. 1 on page 3. This was first published in 'Fire Prevention Science and Technology', September 1977.

PREFACE

It is now thirty years since the Aqua Group was first formed. It arose out of a meeting held at the Talbot Restuarant in London Wall where the price of the three-course dinner was six shillings, or 30p. After one of those meetings five architects and five quantity surveyors got together to put the building world to rights!

We cannot judge whether they have been successful but they did produce *Pre-Contract Practice for Architects and Quantity Surveyors* and *Contract Administration*, in their sixth and fifth editions respectively, and then, more recently, *Tenders and Contracts for Building*, already in its second edition.

There are still five members from the original group and it is one of these, John Oakes, who has led us into new territory with this present volume. *Fire and Building* as a book was born out of a realisation that there was so much to know, so many aspects of the subject from the physical characteristics of fire in buildings to the awful problems of clearing up the mess afterwards and claiming on insurance, that here was a subject that the experience of the Group might usefully pass on to others.

Our problem has been to confine ourselves to prinicples and point the way to further reading, rather than cover the whole immense subject in full detail. Nevertheless, we have been able, within that framework, to deal with it comprehensively to the extent of setting out the various measures of prevention of fire in buildings in the first place, dealing with occupation, the fire certificate and management and security, and finally with insurance, claims and reinstatement. In particular we have tried to relate the parts to each other – thus design will affect insurance premium, and so on.

Our thanks are due particularly to John Oakes and James Williams, who have borne the brunt of the work in producing this book. Fire is a serious subject but an Aqua book would not be complete without Brian Bagnall's sketches which, once again, delight us.

Our thanks also to Julia Burden from our publishers who has attended many of our meetings and helped us as your, the readers', representative.

The Aqua Group
May 1984

The Aqua Group
Tony Brett-Jones CBE, FRICS, FCIArb (Chairman)
Brian Bagnall BArch (L'pool)
Peter Johnson FRICS, FCIArb
Frank Johnstone FRICS
Alfred Lester DipArch, RIBA
John Oakes FRICS, FCIArb
Geoffrey Poole FRIBA, ACIArb
Colin Rice FRICS
John Townsend FRICS, ACIArb
James Williams DA (Edin), FRIBA

INTRODUCTION

The purpose of this book is that of a guide, something of a 'Pilgrim's Progress', through the jungle of innumerable byelaws, regulations and safety measures which must be observed; through the hazard of fire itself to that obscure area of fire insurance, claims and reinstatement.

Very often a designer is uneasily aware of his limited knowledge and his heavy responsibility in matters of safety. But such is the proliferation of documents on the many aspects of the subject that a form of defensive amnesia sets in and he chooses to push on regardless, relying on those in control to step in with guidance and solutions to his design problems before any real damage can be done.

The consequence of fire seems to be the domain of a very few specialists, mostly insurance experts and surveyors acting for the insurance institutions. Faced with instructions to act on behalf of a building owner, to analyse the damage, advise on reinstatement and give guidance in optimising insurance benefit, many designers and quantity surveyors become embarrassingly aware of their shortcomings. What they will undoubtedly learn is that a better understanding of fire and insurance before the event would have made their task very much easier after it.

Legislation, Codes of Practice and the rules governing insurance are constantly changing. Statutory requirements actually differ from area to area. London and many of the large administrative areas have their own byelaws; Scotland's regulations differ from those of England. Even to fire prevention officers, revisionary Acts of Parliament can become bewildering. A definitive volume encompassing the entire subject would be impenetrable and impractical. Like the global issue of monthly *Notices to Mariners* advising on the constant change in navigational waters, an encyclopedia of information on fire would require the constant issue of errata and addenda.

Nevertheless the Aqua Group feels there is a need for there to be set down in a single volume and described in simple terms, all the basic matters about fire which should form part of the essential knowledge held by those involved in building design, construction and maintenance. This book attempts to fulfil that role.

'The eternal triangle'

Chapter 1

UNDERSTANDING FIRE

Loss of life and property by fire most commonly occur because of ignorance about the nature of fire and its behaviour, possibly lethal, in circumstances that normally appear perfectly harmless. If a successful campaign is to be conducted against fire and its dreadful consequences then the adversary must be fully understood. Never forget, prevention is better than cure.

The Nature of Fire

The *Oxford English Dictionary* describes fire as 'the active principal operative in combustion'. Combustion is a chemical process; the oxidation of organic material with the development of heat and light. Organic materials are those constituents of compound substances containing carbon in their molecules (carbon compounds).

To occur at all fire, or in its wider context combustion, relies on the presence of three ingredients – fire's own eternal triangle. The parties to the triangle are oxygen, a combustible material (fuel) and an appropriate level of heat.

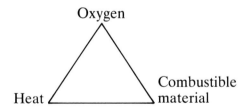

Provided there exists a combination of these three in the appropriate relationship, combustion will occur. The balance of any two may be correct and it requires only the introduction of the third for the process to begin. A change of molecular structure takes places with the absorption of oxygen, the emission of heat and the unmixing ('clearage') of the chemical bond between the elements in the combustible material e.g. from solid to liquid or more commonly to gas.

In summary thus far and in terms relevant to the designer's concern an initial temperature must be achieved at which a combustible material will ignite provided there is adequate oxygen present.

1

The triangle at work

For very practical reasons the designer must be wise to the workings of the triangle. These are best demonstrated by examples that must seem simple but may be surprising.

Most people are aware that combustion can be prevented by dousing with water. This has the effect not of cutting off the oxygen supply as might be supposed, but of reducing the local temperature to a level below which combustion will not continue. Water hosed into a raging fire will merely evaporate and sustain the oxygen supply. Water sprayed around the perimeter of the fire will prevent its spread. However, water sprayed on to a hay barn or granary can promote organic decomposition, causing the temperature to rise and a build-up of heat which cannot escape. The consequence would be a fire. One is taught that a small fire may be suffocated by throwing a blanket over it. However, it is not necessarily appreciated that this particular action is only successful if suffocation of the fire continues with a cooling off period. Water on a deep fat fire causes momentary loss of oxygen but the fat, at a very high temperature, instantly evaporates the water and scatters fine droplets which explosively re-ignite. On the other hand foam or an asbestos blanket is not thrown off by the boiling fat, which gradually cools. In a different context a fire can be starved by isolation from the heat source as occurs when the power supply is cut off from an overloaded wiring system.

Separation of one or other of the members of the triangle is the stuff of fire fighting.

The evidence and progress of fire

It can be said that the very process of combustion generates the ideal conditions for its own development into full active fire. In fact it is only in something like the striking of a match that there is immediate and apparent fire in the form of flame. The usual evidence of combustion in its various stages is as follows (but elaborated upon below):

- Smouldering (non-flaming combustion) – combustion without flame, sometimes with incandescence and usually with the emission of smoke.

- Smoke – clouds of fine particles, usually very hot, the product of incomplete combustion.

- Incandescence (glowing heat) – glowing with high temperature.

- Flaming – oxidation of liberated gases generating heat and light.

Figure 1 illustrates the normal progress of combustion in all its phases in a single place and under ideal circumstances, that is when there is an appropriate fuel, a rising and maintained level of heat and a steady supply of oxygen – all the circumstances to be found in an airy building filled with combustible furniture but having no fire protection. When a fire rages uninhibited through a building the physical process illustrated in fig. 1 progresses through every part of the fabric, rather like a wave unrolling along a beach and then gradually receding.

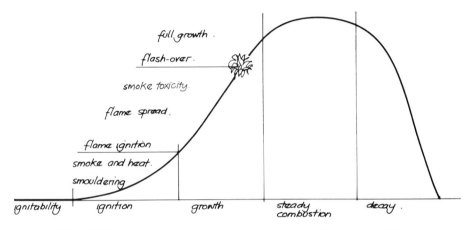

Fig. 1 The normal progress of combustion – the wave of fire.

To begin with then, a rise of temperature from any cause (for example human activity, heat from a chemical action, space heating or electrical overloading) may lead to the earliest stage of fire – smouldering or non-flaming combustion. Both these terms refer to partial combustion at a comparatively low temperature and therefore a low rate, with a limited level of thermal radiation. It is accompanied by thermal decomposition of the fuel material (clearage of the chemical bond as described above) and by the release of combustion products in smoke form, such as carbon monoxide, carbon dioxide concentrates, oxides of nitrogen, hydrogen chloride, hydrogen cyanide and many other possible toxicants depending upon the composition of the fuel.

It should be emphasised at this stage, that some of the combustion products, such as carbon monoxide, can be lethal. Smoke containing noxious fumes and at a temperature of only 80° to 100°C can kill with dreadful speed.

Non-flaming combustion can continue in some materials up to a temperature of 400°C without reaching its flame ignition temperature (the temperature at which flaming combustion commences). However, at the low end of the scale, the flame ignition of timber, is between 250° and 300°C. Many of the combustion products released and accumulated may themselves be combustible given a further increase in temperature. Smouldering combustion at very high temperatures can cause incandescence of the fuel material at the point of combustion, the glowing heat of which might act as an ignition source affecting volatile gases amongst the combustion products already released. This is particularly likely in the event of the sudden admission of air to replenish the hitherto depleted supply of oxygen.

The development of fire is dependent upon the continuing supply of oxygen. In the first place there is nearly always sufficient oxygen present for some degree of combustion to be initiated even if it is only within the fabric of the fuel material itself. With adequate oxygen present a non-flaming fire will develop rapidly, emitting heat and building up the temperature required to make the transition to flaming fire. In a poorly ventilated space in which combustion is rapidly consuming the oxygen in the atmosphere, it only requires a depletion by between 10% and 15% for oxygen starvation to occur. In an atmosphere thus debased the growth of combustion is inhibited and only vitiated burning (or partial combustion) can continue. However, vitiated burning does release further gas or liquid and volatile materials amongst its other combustion products.

The development of fire (fig. 2(A)) differs from situation to situation. In the open air, development is rapid because of the unlimited supply of oxygen and the ready dispersal of combustion products. A fire in a large industrial space is quite a different matter, for although there may be an ample supply of oxygen the combustion products may be trapped and suddenly ignite with the build up of heat. In a series of small spaces, as might be found in a house, the limited supply of oxygen is more likely to cause a vitiated burning with a greater build up of smoke. It is in this situation that injury or death caused by smoke containing toxic vapours (toxic smoke) is most likely to occur.

Flash-over and smoke explosion

Because of the reliance of a fire upon oxygen supply, partial combustion with the production of toxic smoke is a phenomenon generally associated with small and often concealed spaces such as cupboards, ducts, ceiling voids and rooms of domestic scale. Where partial combustion occurs in large spaces such as warehouses, factories or shopping malls, it is probably because the combustion material has not achieved the temperature level at which the transition can be made from developing to flaming fire. In either case this transition can be made suddenly and over a large area and is known as a flash-over.

In broad terms, flash-over occurs when an ordinarily developing fire, or in certain circumstances a smouldering fire, reaches a level of activity at which previously released volatile combustion gases will suddenly

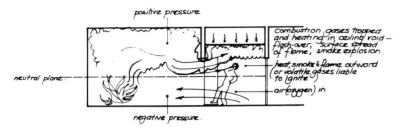

A. Developing Fire.

B. Movement & Statification of Smoke & Gas.

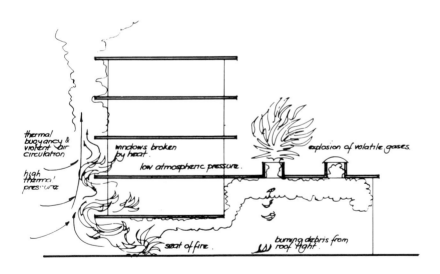

C. Building Form & Spread of Fire.

Fig. 2 Behaviour of fire

ignite and produce flaming across the surfaces of adjoining pre-heated combustible materials.

This very rapid spread of flame differs somewhat from another form of flash-over known as smoke explosion. That occurs when there is oxygen starvation which results in only partial combustion or a smouldering fire at deceptively low temperature. The fire leads to the formation of highly volatile vapours which, mixed with a mist of low volatility materials, is regarded as smoke and in the vitiated atmosphere may be of sufficient concentration to be above the upper explodeable limit without actually igniting. Provided the source of ignition still exists, the introduction of oxygen will bring these products into the flammable range and sudden explosive flash-over will occur.

The typical fire

An everyday situation is easy to imagine. The incorrect wiring of electrical equipment in a small room causes severe overheating of a cable. The heat causes decomposition and combustion of the surrounding materials and heat emission causes flame ignition of the immediately adjoining fabric. The ignition and early burning stages of the fire rapidly consume the oxygen in the atmosphere so that after a while only partial combustion is taking place. This smouldering fire may be of quite a low temperature but producing highly volatile gases. Or it may be emitting considerable heat but producing only moderately volatile gases. Either way these smoke products are gathering at high level in the room and when the presence of the fire is recognised, and a door opened, the sudden inrush of air causes an explosive flash-over which may produce pressures in the order of 5-10 kN/m^2 and which may easily blow out windows and fan lights. Whilst the fire in the room is now on full growth with a steady supply of oxygen, flames, heat and high temperature combustion products are being emitted at high level through the fanlights and traversing the ceiling surfaces in the corridor. Finishes to these surfaces start to decompose and produce their own volatile smoke whilst the base materials begin to heat up until the whole reaches a temperature level at which flame ignition leaps across the affected areas (fig. 2).

As the fire spreads, the supply of oxygen becomes a critical factor. Huge quantities of searing hot and volatile smoke build up in areas where there continue to be flames. The heat and internal pressure which this causes shatter the windows. The smoke explodes into the open air and erupts in flame up the face of the building. Glass to windows in floors above shatters and the swirling circulation of the air caused by the fire sucks the flames into the rooms thus exposed. With ignition, fire growth and steady combustion, the great wave of fire runs through the building. It could be some considerable time before decay of the fire sets in and the wave recedes; unless of course the building is designed to inhibit or contain the fire, or is one in which the fire can be successfully fought.

The spread of fire

Understanding the process of combustion helps in understanding how fire spreads. But whilst one tends to think in general terms of fire as the hazard and destructive force, it is more accurate and useful to think in the separate terms of heat, smoke and flame; each in their own way harmful to materials, injurious to life and to be restricted and fought.

Those who design and manage buildings must constantly remember the three basic media in which smoke, heat and flames are transmitted.

- Building contents – furniture, fittings and fabrics
- Building fabric – finishes, substrate and structure
- Building spaces – concealed spaces, circulation spaces, communicating spaces – any space where smoke can gather

Building contents

There is no doubt that the major fuel for fires is provided by the building contents; it is with the contents that most fires start, as statistics confirm. Whilst designers are usually aware that the furniture and fittings they specify need to be flame resistant, those responsible for management of buildings are bound to have great difficulty in exercising control over users, over what they bring into the building, the changes they make and the way they behave. The main problem with furniture and fittings lies in the intense burning characteristics which they possess and the rapidity with which combustion products, smoke and toxic gases are released. It is these, particularly in the home, which are the main cause of the majority of fatalities from fire in the United Kingdom. Curtains and some types of carpet also contribute to the rapid spread of flame unless treated, for they have a more direct contact with building finishes and structures and readily become involved in the flash-over situation.

The greatest risk, however, lies in bedding and upholstery. Previously, natural materials were used for these – such as wool, cotton, kapok animal hair and latex rubber. These were readily subject to smouldering combustion, often started by cigarettes and not noticed for some time. Modern materials, such as acrylic and polypropylene fabrics, polyester and acrylic paddings and polyether and polyurethane foams, are all less vulnerable to smouldering and require flame ignition. The process can be complicated, however. The action of heat, even quite modest temperatures of about 400°C., can cause clearage and release of volatile fragments in smoke form. With a further increase in temperature, these can reach flame ignition, producing more heat which feeds combustion of the polymer fuel. Once ignited, polymers burn more rapidly and dangerously than natural materials. Covering materials decompose more quickly exposing combustible interiors. At temperatures between 400° and 700°C., the greatest quantity and variety of dangerous combustion products are released. The toxic characteristics of these products may have sublethal effects on human beings, impairing their physical and mental performance and so spoiling their ability to escape. At the other extreme, however, exposure to hot toxic gases produced by polymers can cause collapse and death within a few moments.

The hot gases and smoke produced by the combustion of building contents quickly spread far beyond the area of origin and create the situation in which flash-over or smoke explosion will occur.

Building fabric

Spread of flame over internal finishes and building materials is classified for the purposes of administering the Building Regulations, as described in chapter 3 (fig. 12). The vulnerability of a material to ignition is dependent upon its chemical composition and its surface exposure to air or, put another way, the ratio of surface area to volume. For example, wood shavings will ignite and burn more readily than the same weight of a similar wood in a single block. It follows then that a combustible material used as a finish is less at risk if it is in a dense and smooth form, rather than cellular or fibrous. Man-made finishes are designed to cope with this situation. Many popular natural finishes enjoy no such benefit. Low density timbers such as Western Red Cedar or natural fibre insulation boards have a far greater spread of flame risk than hardwoods. (The treatment of such material is mentioned below and described in chapter 3.)

Among man-made products, plastics pose the greatest risk of spread of flame because of their chemical composition. They fall into two groups:

● Thermoplastic – This group includes polystyrene, polypropylene, polyvinyl chloride and acrylics. Generally they require a high temperature to achieve flame ignition (as was described earlier in the case of furnishings), but are subject to thermal decomposition or change of physical condition on exposure to quite modest levels of heat. As a result they soften and flow at a rate which is dependent on mass and fixing.

● Thermosetting – Examples of this material are polyurethane, polyisocyanurate and the less hazardous phenol formaldehyde. These do not soften but with flaming combustion decompose to leave a shrunken and charred residue, producing larger quantities of flammable or toxic vapours. The residue can form an insulating or protective layer.

Plastics can be produced with varying degrees of resistance to combustion and their characteristics can be exploited even to the extent of having a role in fire protection (see fig. 12). Because of inherent risk, however, the designer must be cautious and fully aware of the performance of the products being specified. For example, polymer-based products exposed in locations where fire might occur, such as light fittings, ceiling finishes and roof lights, once ignited burn easily, produce particularly hazardous toxic gases and melt burning droplets on to furnishings and fittings below. Polyvinyl chloride (PVC) sheathing to electric cables is vulnerable to thermal decomposition when heat is caused by electrical overloading, and will spread smouldering combustion from one compartment to the next at temperatures above the flame ignition level of adjoining materials. PVC pipework in drainage can also spread fire. Burning pipes will lead smoke from compartment to compartment, adding to it their own combustion products. As the pipes burn their way through walls and ducts, the holes left by the collapsed pipes provide a way for smoke, heat and flames to follow behind.

It is interesting to note that, the 'surface spread of flame' performance of materials is not substantially affected by the application of oil-based or polymer paints. These generally deteriorate under heat conditions and expose the material below, burning only once the substrate material itself has ignited and caused a significant rise in temperature. Special flame-retardant treatments (to timber for example) expand under heat and isolate the substrate material from the necessary supply of oxygen.

The introduction of thermal insulation creates problems for the designer. The conservation of heat by the use of heavy insulation adds to the growth rate of a developing fire. In a confined space with an adequate air supply, rapid fire growth will actually reduce the total period of combustion. Insulating material can prevent the transfer of heat to unexposed surfaces over which the spread of flame might otherwise occur and can even prevent a rise of temperature to the combustion level of adjoining materials before the initial fire has burnt itself out. In this way good insulation may prevent the spread of fire to a roof surface. But this of course depends on the insulation material itself being non-combustible (mineral and glass fibres, etc.). Some insulations, such as ureaformaldehyde, phenolformaldehyde and foamed polyurethane chips, are liable to charring, smouldering and finally, at high temperatures, flame ignition. They are notorious for the production of heavy and noxious combustion fumes.

On the other hand, the use of heavy insulation can create a risk by actually preventing the emission of heat, for example from overloaded electric cables. Whilst it is most unlikely that such heat would reach the level at which a combustible insulation would ignite, it is possible for the heat to build up and cause smouldering to spread along the cable until it makes contact with a material having a lower flame ignition temperature. The ventilation needed to prevent condensation in insulated spaces can be the final cause of flame ignition when such circumstances, rare as they might be, occur. Nevertheless, insulation can conceal the spread of smouldering combustion and contribute to flash-over.

The performance of the materials of which roofs are constructed is of importance. The inner linings of the roof are those most directly exposed to the heat, smoke and flames of any fire within. Ceiling linings may fall before igniting, exposing the structure behind, which in its turn may fall whilst burning, spreading flames to furniture, fabrics and floor finishes below. On the other hand, should the roof linings be well secured and non-combustible, or should there be no lining at all, then the roof may confine heat and combustion products, assisting the spread of flame within the building. Roof lights are a particular source of danger since those that melt can produce burning droplets which ignite materials below. In addition, their collapse can cause the release of volatile gases which immediately explode in the air and spread flame across the top surface of the roof.

The outer face of the roof is vulnerable, not only to fire breaking through from below, but to fire or burning debris from an adjoining building or a distant part of the same building. Surface spread of flame on the top face of a roof is unrestricted by lack of air. Ignited material such as asphalte on a flat roof will not melt and run away but, sustained by the build up of heat underneath, will eventually burn, disgorging combustion products into the atmosphere and exposing the surrounding areas to smoke hazards. Bitumen burns easily. Either might be disastrous if the roof forms an escape route, and particularly so should it be on a multi-storey building where rescue could depend on the use of helicopters. Collapse of the roof membrane will cause the burning surface materials to become added fuel for the fire below, the accumulated heat of which might cause smoke explosion. The total effect will be such a build up of heat as to endanger the primary structure of the building, be it brick, steel or concrete.

Of all the constraints, those associated with fire are probably the least considered when building elevations are being designed. Yet it can be the outer face of the building that offers the most direct passage for flames from one part of a building to another. Two factors are involved.

The first is the response of the cladding materials to heat and flame; the second is the ability of flames to reach from one opening to another, regardless of the combustion characteristics of the intervening materials.

It is ironic that, by tradition, low buildings are generally faced in safe and non-combustible materials such as brick and stone (timber would not be used where spread of flame might be a critical factor) whilst modern construction methods used for multi-storey buildings foster the use of cladding materials which are liable to burn or fail in their mechanical performance upon being subjected to high temperatures. In most incidents fire reaches the face of a building from its own interior by shattering the glass in windows. The window openings then behave as horizontal flue outlets disgorging heat and combustion products up the face of the building. The sudden and plentiful supply of oxygen can, as has already been described, cause sudden eruptions of flaming combustion across the face of the materials above and to some degree to either side of the opening. It is then that the response of the materials involved becomes relevant.

Glass in adjoining windows, panels and curtain wall systems will shatter, scattering lethal debris on people below, leaving insulation exposed to add fuel to the fire whilst glazing mullions and transomes buckle under the heat. In the same circumstances, glass fibre reinforced plastic panels may distort, melt and at high temperatures burn with ferocious activity, spreading flame across the face of the building. Recent developments in these materials have diminished the risk of surface spread of flame whilst glass fibre reinforcing and the use of good mechanical fixings prolong the resistance to collapse. Nevertheless, polymer materials are essentially combustible and the fearful lesson of Summerland referred to in chapter 7 remains with us.

Concrete cladding panels too are liable to failure when subjected to fierce heat. High temperatures will build up, causing expansion of the steel reinforcing and the distortion of metal fixings. The concrete will spall and split (thin panels will even buckle) fixings will fail and the cladding will fall readily.

The ability of tongues of flames to reach from one opening to another might be encouraged by combustible cladding materials, but is hardly deterred by the use of non-combustible ones. The greatest deterrent lies in increasing the distance between the openings themselves. Although this might appear to be achieved by having a strongly profiled facade, for example by floors or fins projecting forward of the glazing line, the fact is that such profiles actually turn flames towards the glazing, due to thermal buoyancy creating its own violent pattern of air circulation within partially enclosed spaces (fig. 2(C)).

The beginning of the end for a building on fire might be said to come with the distortion of the main frame by heat. The familiar after fire picture of the steel frame distorted almost beyond recognition tells a story of extraordinary thermal expansion and of forces that can demolish brick walls and concrete floors even before the fire has reached its peak.

Concrete structures are not immune to such movement. It is possible that the combined forces of differential thermal movement between steel reinforcement and concrete within a structural member and of the massive dead loads imposed upon it by the structure itself can lead to ultimate collapse. In a very heavy composite structure of steel framing encased in concrete, heat and thermal movement may cause the destruction of the concrete casing but the steel frame will remain to carry the deadload, deferring collapse until it too fails by buckling.

In some ways, substantial timber frames may present less risk of building collapse. Although there may be a degree of surface spread of flame over columns and beams (unless they are specially treated) the charcoal residue of combustion acts as a protective layer to the structural core. Thermal shrinkage in timber will cause less damage to non-combustible wall and floor membranes than thermal expansion in the case of steel. The hazard remains, of course, that the surface can transmit flame from one part of the building fabric to another and transmit fire to spaces that are otherwise separated.

Building spaces

With fuel provided in the form of the building's contents and fabrics, fire needs spaces for the full exercise of its invasive powers. Any spaces in which heat and smoke can be harboured and built up, be they concealed or obvious, small or voluminous, contiguous, communicating or merely leaking are all significant stepping stones for developing fire.

The presence of an active fire in a building is heralded by advancing smoke. Hot smoke and gases have a thermal buoyancy which, in a confined space, builds up atmospheric pressure above the ambient. For example in a duct or lobby this pressure will cause smoke and gases to squirt through small apertures, cracks round doors, melted pipe holes, etc., thus transmitting the hazard from one space to the next. In a large space a micro-climatic condition occurs, in which a relatively high pressure zone of hot smoke and gas creates air circulation, propelling lethal clouds vertically and horizontally into the uncontaminated lower pressure zones. Hot advancing smoke does not drift, it races.

Essentially any building is an assembly of spaces which can be identified as having either a horizontal or a vertical emphasis.

A vertical space, such as a staircase, a lift well or undivided duct, might be regarded as a flue gathering unburnt combustion products and heat and transmitting fire from one level to another by means of convection – the natural upward movement caused by thermal buoyancy. The flue effect can be destroyed in vertical ducts by a separation at each floor. In lifts and staircases this is impractical and it is more important to exhaust the pressure and smoke that otherwise might gather in them.

Bearing in mind how smoke can be propelled by the differential pressures that develop, a horizontal space such as a corridor or ceiling void can also be regarded as a flue. But because it is more of a trap of heat and volatile gases, it becomes a flue in which flash-over, or smoke explosion, can occur. As in the case of vertical flues, subdivision by fire-resisting doors or compartmentation can destroy the flue effect. Nevertheless people have to move through such spaces and the subdivision might not be very effective. The subdivided spaces are certainly useful in providing safety zones on escape routes but at the same time can be impossible to exhaust, except by mechanical means, should they become filled with smoke. Thus there tends to be a dramatic pace in the spread of fire through horizontal spaces – the reason why, in multi-storey buildings, fire will usually spread through the whole of one floor before moving to the next.

The most dangerous of the horizontal spaces to be found in any building are the voids above suspended ceilings. The reasons are many and vital.

- They are the natural trap for hot combustion products.

- They usually contain electrical and other services which may be a source of fire.

- They are frequently formed of combustible materials and contain combustible insulation.

- If insulated, they encourage the rapid build up of heat.

- They link the extremities of large spaces and may even link adjoining spaces via service outlets.

- They present ideal conditions for flash-over to occur.

- They are concealed – what goes on is out of sight until the ceiling collapses.

The worst aspect of fire in a suspended ceiling is the unimaginable speed with which it develops and the terrible harm caused by the ceiling collapse. Unfortunately, suspended ceilings are frequently added inexpertly by owners as an afterthought, usually with no knowledge of the potential fire risk. Designers, too, are frequently tempted to specify inexpensive but dangerous materials, where risks ought never to be taken.

In very large spaces, stratification of smoke and gas occurs as hot combustion products rise to the ceiling and expand sideways, cooling as they go until they reach the furthest walls where they fall and are drawn back towards the fire as it sucks in air. The result is layers of smoke moving in opposite directions and at different temperatures. On several occasions in large shopping malls, people have been confused by the movement of smoke and, in an effort to escape, have fled towards the base of the fire. It must not be thought that because such spaces are large and the smoke trapped within them has the opportunity to cool that they are safe from the danger of smoke explosion. This is far from the truth. Large spaces tend to contain large quantities of potential fuel which, even in a vitiated atmosphere, might burn, producing tremendous heat energy. Recently, when a fire in a shopping centre spread to a model shop, the burning of the polystyrene packing round the model kits discharged hot gases into the smoke-logged mall and a peak heat output of 26 megawatts was recorded. The sprinkler system was capable of containing a fire with no more than 5 megawatts output. Smoke explosion in such circumstances would blast heat and flame into every corner, turning a terrifying smoke trap into an inferno. Enlarge the scale of such an event to that of a massive factory or warehouse in which dangerous products are stored and the result could be a blazing holocaust that might affect a whole community.

'*Special plant*'

Chapter 2

ASSEMBLING THE FACTS
AND COUNTING THE COST

The designer's responsibility

The statutory responsibility for complying with fire precaution rules and for providing and maintaining means of escape rests with the owners and occupiers of buildings, but the designer must never lose sight of the fact that he bears a heavy responsibility in these matters. His brief from his client will, or should, contain comprehensive instructions as to the client's requirements but, though it may touch upon matters of safety, it is unlikely to deal with fire prevention, fire fighting and means of escape in any depth. These aspects of the building must be identified, researched and brought into the design on the initiative of the designer. A designer who fails to take that initiative might well be guilty of negligence.

This initiative regarding fire regulations must be taken whilst the brief is being compiled. Not only will compliance with the fire regulations influence the design and layout of the building, it will also have an impact on the cost – not on just the capital cost, which will probably be foremost in the client's mind at this stage, but also on the subsequent cost of the building in use, particularly the cost of insurance.

Assembling the facts

The designer will normally be expected to have a reasonable knowledge of the provisions to be made for means of escape and protection, but it is unlikely that he will have at his fingertips the great mass of details embodied in the controlling Acts, byelaws and Codes of Practice, all of which can have a profound effect upon the location, design and cost of the building. He must therefore gather together all the relevant legislation and regulations and relate these to the building which he is about to design. If he fails to do this right at the start, much abortive design work and waste of time and money are likely to be incurred.

So long as the specific use to which the building is to be put is known to the designer, the full requirements of the relevant legislation can and must be applied. However the ultimate use or user may not be known, as for example in the case of large multi-purpose building complexes for public assembly or, more commonly, speculative commercial and industrial developments. Then only general requirements can be met in advance, additional provisions having to be made at the time of occupation or subsequent alteration. In such situations the designer must be aware of the need to allow for flexibility in the design of his building with opportunities to extend circulation, or make provision for ventilation and special plant.

At the outset of the design process the designer must not only acquaint himself with the relevant Acts, regulations and Codes of Practice, but should also consult on general principles with the fire prevention officer who will be responsible for inspecting and certifying the building on completion. More detailed points of construction and other matters covered by Building Regulations should be taken up with the local authority's building control officer and, if necessary, with the local licensing officer.

In the case of buildings where there is a possibility that special risks may be involved or might arise in the future, there should be early contact with the owner's or future occupier's insurance company. This will enable the insurer's representatives to identify the risks and to offer advice and also to indicate at an early stage any special requirements the insurers may have.

Present and future legislation

Legislation affecting fire protection varies in different parts of the United Kingdom and is summarised in detail in appendices 1 - 4 under the following headings:–

Appendix 1 Legislation for England and Wales

- General legislation

- Local Acts

- Fire related legislation

Appendix 2 Legislation for Inner London

Appendix 3 Legislation for Scotland

- General legislation

- Local Acts

- Fire related legislation

Appendix 4 Legislation for Northern Ireland

- General Legislation

- Fire related legislation

Experience and research lead to a constant process of amendment to existing Codes and Regulations, the introduction of new legislation and the consolidation of fragmented but proven standards.

Much of the legislation under the headings above is, at the time of writing, under review. The Building Regulations 1976 – England and Wales, to which frequent reference is made in this and subsequent chapters, is due to be replaced with new legislation based substantially upon Codes of Practice and embodying many of the requirements of local Acts, most notably those of Inner London.

London's first building byelaws were introduced in 1199, but it is of course to the catastrophe of the Great Fire in 1666 that Inner London (as it now is) owes its heritage of fire-related legislation and current exemplary Codes of Practice. Even if the legislation has become unduly complex the codes are universally acknowledged for their high standards and practicability and undoubtedly provide a frame of reference upon which future national standards may be based. At the present time the Greater London Council is the fire authority for Greater London and area administration is in the hands of the District Surveyors.

In the remainder of London the Building Regulations apply and control is exercised by the building control officers of the London Boroughs.

The form of administration in Northern Ireland and the different legal system in Scotland mean that common regulations cannot be used throughout the United Kingdom. Although tending to follow the English and Welsh patterns, Northern Ireland and particularly Scotland will continue in the administration of their own Acts of Parliament. There is a degree of simplicity in Scottish legislation in that there is a clear difference between the regulations concerning the design, construction and alteration of buildings on the one hand and the particular requirements of a limited number of buildings in use on the other hand.

Codes of Practice and guidance

Outside the legislation, but with considerable and increasing influence, British Standards and other Codes of Practice set up principles of design and construction which provide the designer with the choice of alternatives in which he needs to balance risk against cost. The most important of these guidance publications are as follows:

● GLC Code of Practice:
  Means of Escape in Case of Fire, revised 1976

● GLC Code of Practice:
  Means of Escape, Houses in Multiple Occupation, 1978

19

- BSCP 3 chapter 4:
  Precautions against Fire
  Part 1: Flats and Maisonettes
  Part 2: Shops and Department Stores
  Part 3: Offices

- The Home Office Guides to the Fire Precautions Act 1971
  1 Hotels and Boarding Houses
  2 Factories
  3 Offices and Shops

- Construction of Buildings in London (the London Building Acts)

- Current publications from the Building Research Establishment

The current Building Regulations refer frequently to BS CP 3 but Chapter IV, Part 3 (Offices) has already been withdrawn and replaced with BS 5588 Part 3. This is a foretaste of the new legislation based upon revised codes to which reference has already been made.

Insurance and the Rules of the Fire Offices' Committee

The whole subject of fire insurance, the rules which control it and aspects of design and construction affected by it, are covered fully in chapter 5. Nevertheless, preliminary reference is made here to the subject which is one of the most important to be considered at the design stage.

Most buildings will be insured by the owners against loss or damage by fire. The sum insured and the terms of the insurance will be agreed by the owner and his insurance company. The insurer will require that the buildings comply with the terms of the Building Regulations and other relevant Acts. However, in a great number of cases these are insufficiently specific to provide good reason or grounds for the appropriate insurance. For example the risk of fire and magnitude of possible loss is far greater in a flour mill or warehouse with combustible packings than in a private veterinary surgery or community library, although they are subject to control under the same sets of Building Regulations and general legislation. To differentiate therefore between the various classes of building and the inherent risks, the insurance companies look to an additional set of rules; those of the Fire Offices' Committee (F.O.C.)

Most of the fire insurance companies in Britain, the tariff companies, are members of the Fire Offices' Committee and their rules provide the basis for calculating insurance premium rates in all types of buildings. Indeed so highly regarded are the rules that independent insurers and Lloyds underwriters also accept them for guidance, whilst local authorities and fire prevention officers regard them as an acceptable, but unofficial, supplement to the Building Regulations in those cases in which interpretation or content of the Regulations fail to resolve the needs of safety.

For the designer, too, the rules provide a most useful guide, although it must be remembered that they are subject to constant revision in order to keep them in line with the latest regulations, research findings and Codes of Practice to which they make frequent reference.

Familiarity with the F.O.C. Rules, as with all relevant legislation, regulations and Codes of Practice, is essential for the design team as they will influence design, cost and long term viability of a project from the start.

Balancing the cost

At first glance it might seem that very little control can be exercised over the cost of meeting the requirements of the multitude of fire safety regulations applicable to a particular building project. This is far from true although it must be said that success lies, in fact, in making the first decision the correct one. For example, an experienced London developer might be heard talking of a site and proposed development requiring a 'section 20 building'. In fact he is identifying the kind of building as one subject to the onerous conditions of section 20 of the London Building Acts which is concerned with the special provisions to be made in commercial buildings in London over a certain height or volume (see appendix 2). The cost of such provisions may seem high in normal circumstances. Seen in the light of prevailing land values, overall development costs and final profitability, they will perhaps appear as an acceptable part of the whole investment; certainly a reasonable balance of cost against risk. With one eye on the regulations and fire safety, a decision as to whether or not to build is first made.

Fundamental design decisions then follow. It is a general rule that the larger the size of compartment, the greater is the hazard, the higher will be the standard of fire resistance required both for the compartment and the building structure, the greater the cost and the more onerous the insurance premiums. Decisions as to the building form may hinge on the most favourable possibility for compartmentation. For the most part the maximum size of compartment is fixed, as is set out for England and Wales in chapter 3. In many situations, such as shopping centres where large spaces are necessary, the maximum size of the compartment can be doubled provided that sprinklers are installed. This clearly has ensuing financial implications.

So it is that many of the principles of fire safety, described in chapter 3 as the Passive Measures and identified by reference to the byelaws together with the Codes of Practice, offer the designer a range of choices which will have a bearing on the economics of the building, for example:

- Use of the site – the manner in which the permitted density of development is achieved; the building's proximity to the site boundary and the impact upon compartmentation, window openings, daylighting; the consequent availability of useful or lettable floor areas and the resultant return on investment.

- Building height – affects compartmentation, the fire rating of both structure and enclosure, escape routes, lifts, and fire brigade access; all factors constituting major constraint on economic viability.

- Multiple or separate tenancies – the economic use of floor space; the need to produce lettable floor areas, provide party walls, shared or separate entrance, numbers of staircases.

- The number and width of escape routes – related to building use; affects maximum number of people who may use building (or floor), the subdivision of floors, travel distance, dead ends and as a consequence the whole floor shape.

- Choice of a structure – heavy- or light-weight, wet or dry; traditional and economic or high speed and costly with early return on investment; fire rating a factor.

- Design of elevations and fire rating of external materials – surface spread of flame, location and size of window openings, cill heights, degree of natural ventilation, thermal performance; all matters relevant to speed of erection and economics.

- Roof construction – form and choice of materials, their resistance to fire and spread of flame; proximity to adjoining hazards, access and maintenance, thermal performance, condensation.

- Surveillance, detection and alarm systems – with economic advantage can affect the provision for means of escape and separation of fire risk areas, etc.

- Automatic sprinklers and other prevention systems – costly provisions that may be necessary to the use of the building but might otherwise allow economy in compartmentation and greater flexibility in the use of space.

- Smoke exhaust systems – may be necessary for certain public buildings but otherwise can affect the building form, the number and design of escape routes and staircases, all with overall economic advantage.

These are but some of the broad areas in which cost effective decisions will be made and, while the cost implications resulting from an analysis of the legislation for new buildings are difficult enough, the problems of cost comparison and choice between different methods of creating fire safety in refurbishment are even more complex.

Refurbishment

The most complex of tasks involved in refurbishing a building will be to bring compliance with fire regulations and means of escape up to date and to house the plant, machinery and electrical installation safely and out of sight; all to modern standards. It is this aspect, rather than the cost of repairing or modernising the shell, which is usually so expensive and responsible for making the total cost of refurbishment almost as much, sometimes more, than the cost of an equivalent new building.

However other factors usually come into play in the decision on whether to rebuild or not. In the case of historic or listed buildings, or to meet conservation area requirements, planning considerations take precedence over pure economics.

With historic buildings there is a body of experience continually devising methods of improving standards of fire resistance without destroying important features. Not all of them are successful and mostly it is a compromise where the most that can be achieved is a reasonable standard – better than it was before. Parts that are totally rebuilt can, by careful design, comply with modern standards whilst building control officers can, where common sense prevails, be helpful by suggesting relaxations of the requirements of the regulations. Early consultation is the keynote to success in this respect.

Priority of facts relevant to legislation and insurance

Whether it is in referring to legislation or conferring with design specialists, it is helpful to have first marshalled into some kind of order of priority all those facts about the building which may seem relevant to fire protection, statutory control and insurance. It will follow that as work or consultation proceeds many points will emerge as additions to the brief, influencing the design as it is developed. It may not be possible to anticipate every facet of a building which might have such relevance, particularly as far as insurance is concerned. However the designer might bear in mind many of the items in the following list:

- Establish Purpose and Occupancy Group so as to identify relevant legislation – see appendix 1,2,3, or 4

England and Wales:
(i) Private dwelling houses, including detached garages and carports with floor areas of 40 m^2 or less.

   (ii) Institutional homes,
       hospitals or schools

  (iii) Other residences, including flats and
       maisonettes not in groups (i) and (ii).

  (iv) Offices.

   (v) Shops.

  (vi) Factories.

 (vii) Places of assembly, public or private.

(viii) Storage and general purposes not
       included in groups (i) – (vii).

Inner London:
   (i) Warehouse, or for trade or manu-
       facture.

  (ii) Office/dwelling or not referred to in
       another class (but not public build-
       ings).

  (iii) Class I use in building shared with
       Class II use, except dwelling.

  (iv) Class I use in building shared with
       dwelling use.

   (v) Accommodating high voltage power
       transformers or electrical switchgear
       or similar risk.

  (vi) Accommodating or displaying petrol
       driven vehicles (the part so used).

Scotland:

A   Residential (with four subgroups).

A1   Houses of one or two storeys but no
      flats.

A2   Houses over two storeys and all flats.

A3   Clubs, schools, colleges, hostels,
      hotels, etc.

A4   Institutions for children and old
      persons, hospitals and nursing homes.

B   Commercial (with two subgroups).

B1   Offices.

B2   Shops

C   Assembly (with three subgroups).

D   Industrial (with three subgroups).

E   Storage (with two subgroups).

● Location of building: and associated risks

Proximity of site to other hazardous uses.

Proximity of building(s) to boundary.

Exposure hazard (transfer of fire from other buildings, etc).

Fire resistance of enclosing walls.

Area and nature of unprotected areas (windows, etc).

Access (fire brigade, etc.)

● Size of building: regulations and circumstances requiring special provision

Permitted height, area or volume.

Compartmentation.

Fire resistance of building fabric.

Number of occupants.

Means of escape, staircase, final exists, etc.

Fire load (factor of use and size).

Smoke control.

Reach of fire appliances.

● Construction: influencing factors and areas of decision

Temporary or permanent structure.

Requirement for protection of structure based on single- or multi-storey building, use of building and public or private occupancy.

Fire resistance of compartment walls, and floors.

Fire rating of envelope, protected and unprotected areas.

Spread of flame over building face, materials and effect of profile.

25

Protection of means of escape, doors, staircase enclosure, lobbies, etc.

Protection of lift shafts, service shafts, ducts, etc.

Division of floors and ceiling voids

Roof finishes, fire resistance and spread of flame.

Roof lights, distance from boundary, fire rating, destructibility, ventilation.

Flammability of insulation.

● Internal planning: the brief and its consequences

Known or speculative use.

Optimum area of occupancy (residential, recreational, educational, etc.) or work space (industrial, commercial) affecting:

    direct distance or travel distance;

    numbers of exits, staircases, etc;

    number of users affecting staircase and corridor widths;

    positions of final exits.

Location of compartment walls and sizes of openings through them.

Locations and numbers of lifts (including firemen's lift).

Ducts and serviceways – location, access and protection.

● Special storage facilities: risks and provisions

Identification of hazardous materials.

Risk of fire and explosion – structural containment.

Temperature control.

Special signposting.

Provision for safe handling, cleaning, etc.

Isolation of drainage.

Removal of noxious combustion products – gas, smoke, etc.

Special fire safety installations – sprinklers, foam, etc.

Artificial ventilation.

Rapid evacuation.

Access to storage area for ambulances, fire fighting appliances, etc.

- Special parking requirements: impact of planning on services and safety

Numbers of vehicles.

Location – roof, basement, etc.

Fire compartments.

Ventilation – artificial or natural.

Sprinklers.

Special lighting requirements.

Special fire fighting equipment – location and type.

Headroom for appliances.

- Fire protection requirements: provision to be made

Alarm systems – audio, visual, etc.

Smoke and heat detectors.

Communication and warning systems – fire and police station.

Fire fighting equipment – extinguishers – location and type.

Dry risers and wet risers.

Sprinkler systems.

Emergency lighting.

Special signposting.

Natural ventilation – lobbies and staircases.

- Mechanical services: degree of isolation and protection

Location and access – roof, basement.

Location and separation from usable space.

Fuel types – location and storage.

Design and structure of flues.

Electrical installations as source of fire.

27

Air handling plant and spread of smoke.

Heating plant and temperature control.

Gas and risk of combustion.

Special piped services – CO_2, oxygen, methane.

Separation of services – gas, electricity, etc.

Lift control and fire lifts.

Lift shafts and ventilation.

This might seem a daunting list but less so once the principles of defence against fire are understood in broad terms. Even then, the extensive legislation can be looked upon as a source of guidance rather than restriction. Counting the cost of building safety and economy is a matter of experience. The message of this chapter is therefore:

● That legislation can mean choice and is therefore capable of cost analysis and comparison.

● That insurance companies can contribute valuable information at an early stage which might effect substantial savings in premiums during the life of the building.

● Economic safety can be achieved by understanding how fire develops and by good design without necessarily implying greater initial cost.

. . . the obvious place in which to look for fire-fighting equipment must turn out to be the place where it is actually to be found . . .'

Chapter 3

MEANS OF DEFENCE –
THE PASSIVE MEASURES

The maintenance of safety of life and protection against loss – the purpose of defence – call for a strategy in design which recognises the nature of fire, the performance of structure and materials and the behaviour of people. The strategy is founded on the principle of instant understanding of the building by those who use it. The designer must think 'fire' at all times. The design must be clear. The fabric of the building must be the safest possible. The apparent ways in and out must indeed be the correct ones. The natural direction in which to escape must turn out to be the safe one. The obvious place in which to look for fire fighting equipment must turn out to be the place where it is actually to be found. Equipment when found must be right for the job. The fireman's natural approach to a fire must turn out to be the best one.

Here and in the following chapter attention is focused on the basic principles governing fire safety, knowledge of which must be as much a part of the designer's stock in trade as is his knowledge of planning and structural requirements. Some detail is included where it will help in the selection of structural systems and materials, where the assembly of components becomes particularly relevant and where some advance knowledge is helpful in briefing specialist suppliers and sub-contractors. Inevitably there must be reference to standard classifications and regulations but the designer must be aware of change, keep abreast of developments and consult with authorities at all stages.

The main defensive measures fall into two broad categories: the 'passive' and 'active' measures. Whilst this chapter is concerned with the passive measures, a description of both is called for at this point for the sake of clarification.

- *Passive measures* – In-built characteristics which are inherently safe and are effective by their presence. For example, clarity of design, good access, simple circulation, good means of escape, compart-mentation, protection of structure, the resistance of materials against spread of flame, ventilation.

- *Active measures* – Those which come into use when fire breaks out, for example detection and alarm systems, sprinkler systems, fusible link doors and shutters, emergency lighting, smoke exhaust.

The building on its site

The earliest consultation with the planning authorities determines the feasibility of a project on the basis of a number of criteria, of which the first will be that of the building use. Not only will it be hoped or required that the building conforms to the designated use for the location, but a judgement will be made within that context as to the real suitability of a particular building to stand amongst its neighbours or to be located in the area in question. Of all the grounds for decision the least obvious is that of danger from fire and explosion, unless of course the proposal is for a building of blatant hazard, such as an explosives factory. Nevertheless hazard lurks in most situations, waiting to be recognised and dealt with according to its potence. Problems associated with industrial installations may come as no surprise. The juxtaposition in cramped conditions, for example, of a filling station and a hotel might create unexpected dangers that are difficult and costly to overcome. Early discussions with the planning authorities should not preclude such matters.

One of the greatest risks to or from a building lies in the exposure hazard, that is, the risk of radiant heat causing fire to spread across intervening space from one building to another or from part to part of the same building. The possible fire hazard that a building presents depends upon its use and size – factors determining its 'fire load' and the intensity of heat that might radiate from it. Fire load is the total amount of combustible material expressed in heat units, or its equivalent weight of wood. (buildings are identified as being of high or low fire load risk; above or below 5 lb per sq. ft respectively when the equivalent weight of wood is used.) The element of risk is then at the heart of the matter in considering:

- The distance between a building and its site boundary or an adjoining building.

- The manner in which the building volume is arranged.

Not unnaturally the site boundary is generally the perimeter of the land under single ownership. However, where a building fronts directly onto a road, the centre line of the road is, for the sake of interpreting the regulations, treated as the boundary.

Separate buildings within a boundary under single ownership must be distanced from each other in the same manner as defined in the regulations for individually owned buildings. If joined together as a single building of similar form, then the whole is considered in relation to the boundary only. The distance from the boundary can be reduced by subdivision or compartmentation of the space within the building so as to break down the effective fire load into smaller parts, limiting the size of a possible fire and the heat that it will radiate. When a building stands on the boundary of ownership then a 'separating' or 'party wall' situation exists. Such is the case in a terrace of houses where each unit becomes in effect a separate fire compartment.

The good behaviour of the properly built terrace, when fire occurs, is well proven. Less predictable is what might happen in the spaces between separate buildings, described as 'undefended spaces' if the construction of the walls of the enclosing buildings or the openings in them, expose the area to fire risk. Burning or collapsing walls reduce the time available for escape and endanger the lives of passers-by and fire fighters. Doors and windows facing each other across spaces can create fire passages. Narrow spaces might behave like flues, rapidly exhausting smoke from within the building and drawing in fresh air to fuel the seat of the fire. A restricted smoke-logged space suddenly exposed to heat might become the centre of smoke explosion and would certainly be a dangerous position from which to fight the fire.

Protection from these risks is inherent in the regulations which we will be examining later in the chapter, but awareness of them is most important at the time when the designer is considering both siting and massing.

Enter the fire brigade

At the beginning of this chapter, reference was made to clarity of access. This is important at all times but most of all when a fire has to be fought. Ready access to the area of a fire, room to manoeuvre the appliances, sufficient depth of space from the face of the building so that ladders of adequate reach can be used, are all essential.

The problem of fire brigade access is greatest in urban areas. Fire brigade vehicles are likely to be delayed and delay gives time for the fire to develop. When the fire brigade arrives the site may be congested with parked vehicles and in any case it might be possible to reach only one face of the building with appliances capable of dealing with the fire.

In confronting these problems it is essential to consult the fire prevention officer. The internal arrangement and indeed the very form of the building might well be influenced by the needs of escape from, and protection of, those parts of the building to which access cannot readily be gained. In siting the building unobstructed access, the 'fire path', for the use of wheeled escapes, turn-table ladders or hydraulic platforms, will need to be provided. Any gradient to such access must be limited and it must be kept free of such obstructions as bollards that cannot be demounted, lamp standards and trees.

The rules for access, published by the London Fire Brigade, provide an invaluable reference for standards acceptable throughout the country. In concise terms these give information on the construction of fire paths, widths and turning circles, maximum gradients, space requirements for ladders reaching various heights of building and provision of hydrants and wet and dry rising mains within the building.

The building form and the Building Regulations

The current byelaws aim to achieve control of fire safety by consideration of four primary aspects of building:

- Purpose
- Size
- Separation or division
- Resistance to fire.

Every aspect of protection is determined by a building's use and the risk associated with it and is the basis of the classification into which it falls – referred to as 'Purpose Groups' (England and Wales) or 'Occupancy Group' (Scotland).

However unfavourably the designer responds to the obscurity of regulatory language, it is usually true that a sense of direction will be achieved by reading the introduction or glancing over the section titles of the appropriate byelaws.

The Building Regulations for England and Wales in Part E (Safety in Fire) chart the rules in purposeful and useful order.

- E2 Designation of Purpose Groups
- E3 Rules for measurement
- E4 Provision of compartment walls and compartment floors
- E5 Fire resistance of elements of structure
- E6 Fire resistance of the fabric and components of the building
  to
  E17

The classification of building types differs for England and Wales, the Greater London area and Scotland. Figure 3 gives the 'Purpose Groups' defined in the Building Regulations for England and Wales, (and restrictions in height and area, volume or 'cubic extent' and numbers of floors relating to each of those groups). Part D of the Building Standards (Scotland) Regulations 1971 to 1979 defines occupancy groups for buildings not dissimilar to those for the Building Regulations for England and Wales. At the time of writing many cities, notably London, have their own Acts giving variance to the basic theme which emphasises the more onerous conditions prevailing in areas of dense urban development. It must be emphasised that the designer should check the application of local byelaws, administered by the building control officer, when working within any major conurbation and not rely solely on reference to the national Building Regulations.

The same points must be made about the regulations controlling unprotected areas (e.g. window openings, etc.), the fire resistance of the enclosure and the distance of a building from its site boundary.

Figure 4 illustrates the 'enclosed rectangle' principle concerning openings in enclosures and the distance between a building and its boundary, as required in the Building Regulations for England and Wales. The unprotected area is expressed as a percentage of the rectangle so described related to the potential fire load risk (i.e. the Purpose Group) and the distance is then read from a table.

Of paramount importance to this set of rules is the fire resistance of the material of the external walls. This may range from one to four hours depending upon the circumstances of location, use and hazard. The regulations, when analysed, will guide the designer towards the correct balance between fire rating and perforation.

Division – compartment and separating walls

Figure 3 shows the limits of volumes for parts of or the whole of a building according to its use. This means that within each class, larger buildings than those shown would require to be subdivided into parts to reduce each volume to the maximum shown in the diagrams and defined in the Regulations. The diagrams also demonstrate that by increasing the compartmentation, the fire rating of the structural elements can themselves be reduced. The smaller the volume, the less is the fire load.

The Building Regulations distinguish between separating walls (party walls) and compartment walls and compartment floors. Separating walls are those which are common to adjoining buildings and are imperforate. Compartment walls and floors are those which subdivide a building for the purposes of separating occupancies within that building or subdivide the building for the purpose of restricting fire. They are also concerned with the fire resistance of structural elements in relation to building volumes.

Compartmented spaces may be linked or, put another way, compartment walls and floors may be perforated. In these circumstances, communicating doors must have a total fire resistance equal to the fire resistance of the compartment wall or floor concerned. The requirements of fire insurance necessitate compliance with the rules of the Fire Offices' Committee and the Committee lays down concise performance specifications for such doors and shutters to which many specialist manufacturers strictly adhere.

Purpose Group I : Residential (small)

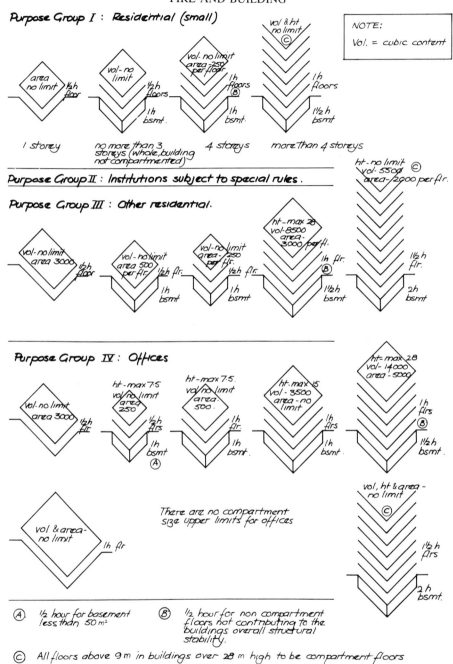

NOTE:

Vol. = cubic content

area no limit ½h floor

1 storey

vol- no limit ½h floors 1h bsmt.

no more than 3 storeys (whole building not compartmented)

vol- no limit area -250 per floor 1h floors Ⓑ 1h bsmt.

4 storeys

vol & ht no limit Ⓒ 1h floors 1½h bsmt.

more than 4 storeys

Purpose Group II : Institutions subject to special rules.

Purpose Group III : Other residential.

ht-no limit Ⓒ vol-5500 area-/2000 per flr.

vol-no limit area 3000 1b h floor

vol-no limit area 500 per flr. ½h flr. 1h bsmt

vol-no limit area-250 per flr. ½h flr. 1h bsmt

ht-max 28 vol-8500 area-3000 per fl. 1h flr. Ⓑ 1½h bsmt

1½h flr. 2h bsmt

Purpose Group IV : Offices

ht=max 28 vol-14000 area-5000

vol-no limit area 3000 ½h fl.

ht-max 7·5 vol/no limit area 250 ½h flrs 1h bsmt Ⓐ

ht-max 7·5 vol/no limit area 500 1h flr. 1h bsmt.

ht-max 15 vol-3500 area-no limit 1h flrs 1h bsmt.

1h flrs Ⓑ 1½h bsmt.

vol & area- no limit 1h flr

There are no compartment size upper limits for offices

vol, ht & area- no limit Ⓒ 1½h flrs 2h bsmt.

Ⓐ ½ hour for basement less than 50 m²

Ⓑ ½ hour for non compartment floors not contributing to the buildings overall structural stability.

Ⓒ All floors above 9m in buildings over 28 m high to be compartment floors

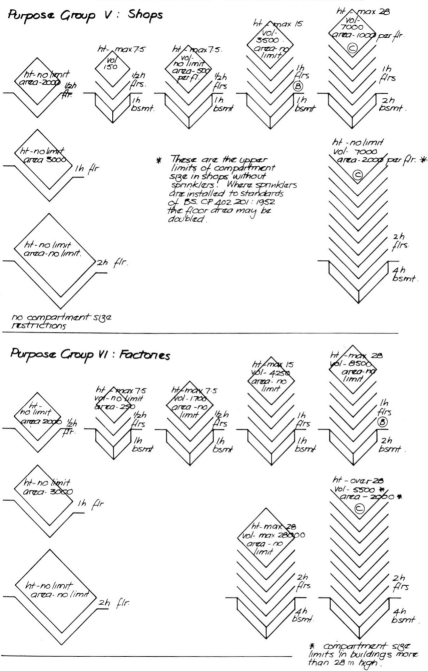

Fig. 3 Building use, size and fire resistance – requirements of the Building Regulations for England and Wales, Part E.

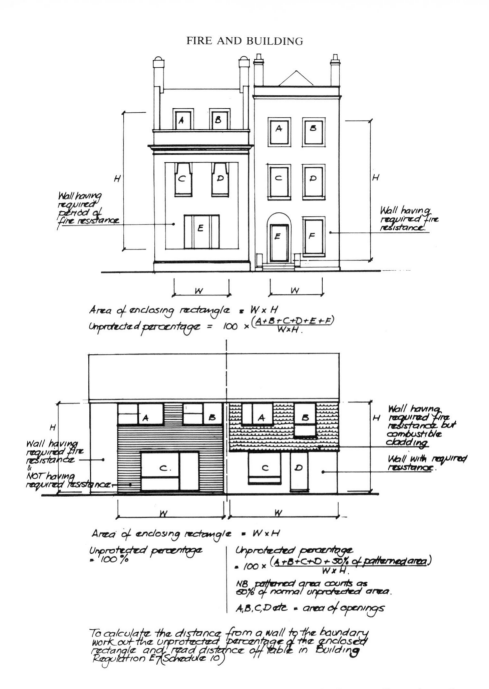

Area of enclosing rectangle = W × H

Unprotected percentage = $100 \times \dfrac{(A+B+C+D+E+F)}{W \times H}$

Fig. 4 The 'enclosed rectangle' principle, to calculate wall to boundary distance.

The characteristics of compartment walls and floors are simple to observe. Their fire resistance will range from one to four hours depending upon circumstances and the most essential requirement is that there should be no combustible material passing through the membrane or bypassing its edges so as to link the spaces on either side. Where it is necessary for services to perforate the membrane these must be carried in non-combustible materials. Small openings must be capable of being sealed, for example by the use of intumescent strips around doors and intumescent paints on louvres. Larger openings such as ventilation trunks will need to be closed with fusible linked dampers.

Clearly the need for compartmentation can have a substantial impact on both the form and the cost of a building, but economy in the long term can be derived from the accrued saving in insurance premiums by reducing the risk of spread of fire. Nevertheless, the requirements of a building may create compartmentation without incurring additional cost. For example in commercial buildings the usual and most desirable form of construction, concrete, automatically creates substantial compartmentation between floors. The same is likely to be true of floors dividing flats or maisonettes although subfloors within maisonettes need have a fire resistance sufficient only to protect the floor from structural collapse before escape has been made by the occupants. By contrast, compartmentation in factories and warehouses can be an enormous problem, running the risk of interrupting production processes or storage methods. Then the designer must weigh up the cost of compartmentation against other alleviating techniques, such as increasing the normal standards of fire detection and protection by the various 'active' measures described in chapter 4.

Safety and the space within

Grappling with the overall concepts of location, massing, structure and expression of purpose, it can be difficult to identify the moment in the design process when an architect first begins to think of the interior arrangement in relation to the fundamental needs of safety.

A well-heralded entrance to a building is a safe blessing; a convoluted approach, a menace. Clear vision of the direction of movement, a sense of location, a natural response to the priority of spaces, and last but not least, easy sight of obviously located stairs and lifts are all ingredients of a building which is good and safe to use. Such unwritten rules must be said to apply to buildings of all classes but particularly, of course, to buildings of public use and assembly, and places where large numbers of people work or rest. In these occur the greatest or most extended volumes of space, wherein lie the most common of dangers – the collection of smoke and incidence of flash-over.

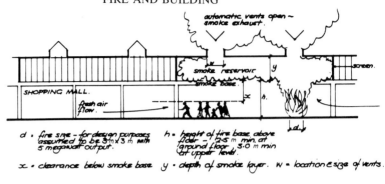

d = fire size - for design purposes   h = height of fire base above
assumed to be 3 m x 3 m with   floor - 2.5 m min at
5 megawatt output.   ground floor, 3.0 m min
at upper level.

x = clearance below smoke base   y = depth of smoke layer.   w = location & size of vents.

A. Smoke Reservoirs - the principles of their design.

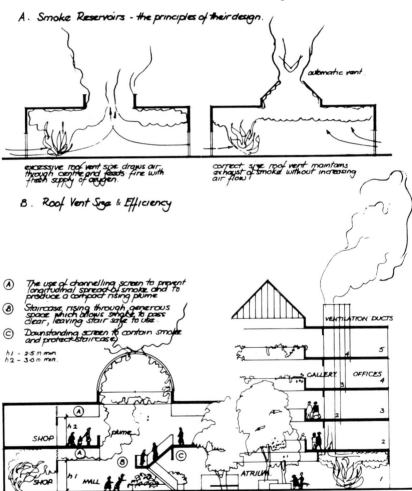

excessive roof vent size draws air   correct size roof vent maintains
through centre and feeds fire with   exhaust of smoke without increasing
fresh supply of oxygen.   air flow.

B. Roof Vent Size & Efficiency

(A) The use of channelling screen to prevent
longitudinal spread of smoke and to
produce a compact rising plume.

(B) Staircase rising through generous
space which allows smoke to pass
clear, leaving stair safe to use.

(C) Downstanding screen to contain smoke
and protect staircase.

h1 - 2.5 m min.
h2 - 3.0 m min.

C. Channelling Screens in a
shopping mall.

D. Protection of an Atrium,
ventilation of each level.

Fig. 5 Smoke control.

Smoke control

The behaviour of smoke is central to the design and location of escape routes and it follows that control of smoke assumes a priority status among the criteria governing the design of large spaces.

When fire occurs and gaseous combustion products of smoke are released, detection, alarm and control are essential with a rapidity which will maximise the opportunity for escape. Detection and alarm are studied later in chapter 4. Control is essentially a matter of natural ventilation or artificial exhaust. Figure 5 illustrates the movement of smoke in a variety of large spaces typical of which are the high and elongated shopping mall, the central atrium of an office block, large exhibition space and factory or warehouse. Attention must be paid to a number of important facts:

- Safety from smoke can depend on the thermal buoyancy of the combustion products. As smoke rises in very high spaces it is liable to cool and fail to set off high level sprinklers or their alarms. Smoke detectors are then essential.

- In medium and low spaces, heat convection sets up air currents that can spread the smoke plume and reverse the direction of smoke as it cools and falls – thus confusing occupants as to the direction of the fire. Smoke movement should therefore be limited by screens forming ventilated smoke reservoirs.

- A plume of smoke from a fire underneath a gallery or low area within a large volume will rapidly spread sideways before spreading upwards into the main space unless baffles are provided. Plumes should be kept compact so as to avoid damage over an extended area and to facilitate rapid ventilation via bypass ducts or exhaust systems.

- The size of smoke reservoirs should be limited so as to minimise the spread of smoke, its cooling and downward mixing with clean air. (For example, 1000 to 1300 m^3 in shopping malls, 2000 m^3 in industrial buildings.)

- Air must be permitted to enter at low level to enable smoke to be exhausted at high level.

- Central spaces place at risk communicating high level circulation areas which must be suitably protected.

- Staircases linking levels within large spaces should be free-standing or have large wells so that smoke may bypass them and leave them in comparative safety.

The form of ventilation which is to be provided might be any of the following:

- Permanent high level natural cross-ventilation.

- Natural ventilation through destructible (by heat) roof lights.

- Fusible link-operated roof vents (active measure, see chapter 4).

- Air pressure systems (active measure, see chapter 4).

Experience with air pressure systems, seemingly the most efficient means of smoke control, is so far rather limited. The passive methods of ventilation are well known and well tried but it must be remembered that any method requires air input at low level which will supply oxygen to the seat of the fire so that the provision of a sprinkler system is a prerequisite. Sprinklers and artificial ventilation systems are active measures with which we will be dealing in chapter 4.

People may survive for quite a time in a large space filling with smoke so long as smoke explosion or flash-over do not occur. This is not so in small spaces like corridors which can fill with terrifying speed so that occupants are immobilised and asphyxiated before making escape. Loss of visibility is the greatest deterrent to escape. Corridors, lobbies and staircases on escape routes must therefore be defended against smoke penetration.

The way out – corridors, lobbies and staircases

In planning internal circulation and locating escape routes, it is essential to keep the escape objectives in their proper sequence in one's mind. In

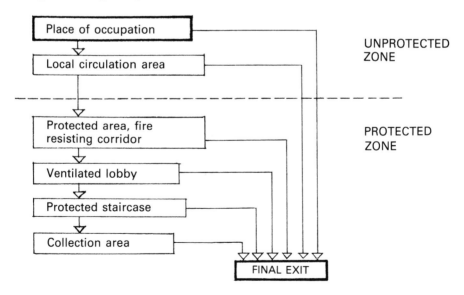

Fig. 6 Sequence of escape objectives.

fig. 6 the areas listed fall into either unprotected or protected zones. Any one of the protected zones can be regarded as a place of safety, albeit temporary, and is important in terms of rescue or partial recovery after escape from the area of immediate fire hazard.

The number of stages intervening between the place of occupation and final escape depend upon circumstances. A single-storey building without complex security problems may have emergency exits from the place of occupation direct to the exterior. Multi-storey buildings of large individual floor areas may have many protected spaces intervening between occupancy and final exit depending on:

- The distance which can be travelled, the 'travel distance', before smoke and heat make movement impossible – e.g. length of corridor (fig. 7).

- The need to isolate communicating escape routes – e.g. ventilated lobbies to escape stairs.

- The need to provide collection points so as to control and phase mass escape from buildings of large occupancy – e.g. protected collection areas (fig. 8).

- The need to protect the point of final escape from adjoining fire risk areas.

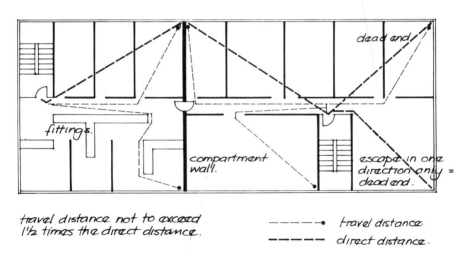

Fig. 7 Direct and travel distance.

Travel distance means the actual distance to be travelled by a person from any point within a floor area to the nearest exit leading to a place of safety, not in a straight line but taking account of walls and fittings. This is not to be confused with 'direct distance' as used in the byelaws,

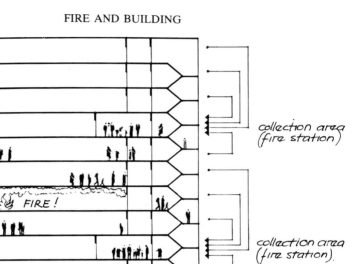

ALARM
STATUS .

alert.

alert.

alert

evacuate.

evacuate.

alert.

alert.

FIRE!

collection area
(fire station)

collection area
(fire station).

In multi-storey buildings
maximum travel from seat of fire
should not exceed five floors up
or down.
Phased evacuation - floor of fire
location and floor above, other floors
follow as necessary .

Fig. 8 Phased alarm and evacuation; collection areas.

(see below and figs 7 and 9). It is based on the somewhat arbitrary assumption that a mobile adult can travel ten to fifteen metres per minute in a smoke filled space where there is some degree of visibility and presence of oxygen, if only at floor level. Similarly it is assumed that the infirm might travel six metres in a minute. In old persons' homes, frequent subdivision of escape routes is necessary to minimise the travel distance and the time taken to reach a safe place such as the next section of an escape corridor. Where non-ambulant occupants are involved and their escape depends upon the strength and mobility of a few, then protected holding points will provide temporary refuge until help arrives or at least they will provide the opportunity for leisurely evacuation. Such spaces will normally form part of the general circulation area, the size of which will determine their usefulness in this role.

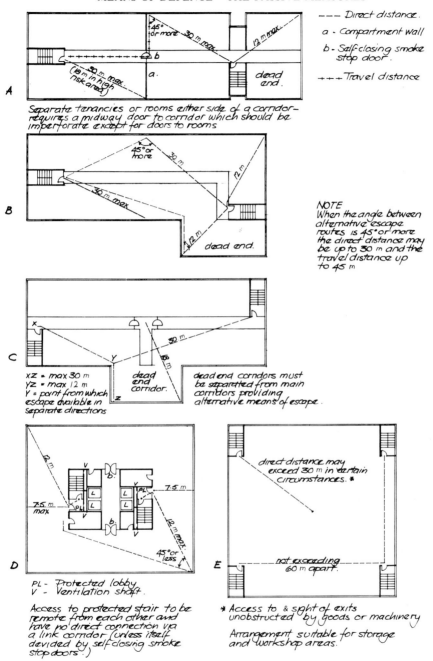

Fig. 9 Location of staircases, 'means of escape' and 'direct distance'.

The location of staircases and their characteristics, either for general usage or primarily as a means of escape, are bound up with economy, the manner in which floors are to be divided among users and last, but by no means least, the travel distance factor. However, travel distance cannot always be identified at the basic planning stage for the exact arrangement of partitions and doors may not immediately be known. For this reason, 'direct distance' is used as the criterion. Figure 9 illustrates a number of staircase dispositions within various building plan forms. The plans are of course only typical of the wide variety of uses and circumstances with which the designer is faced and all of which are subject to particular rules. They are nevertheless sufficient to enable one to appreciate that building economy is to some extent bound up with tackling the criteria related to fire safety in the correct order as illustrated in fig. 10.

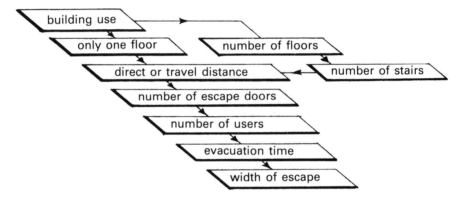

Fig. 10 The criteria governing design for means of escape.

Of great importance are the danger and difficulties that arise from dead end situations, i.e. those from which no choice or alternative means of escape exist. These are plan arrangements into which designers may be tempted in pursuit of economy. (fig. 9).

A whole floor must be regarded as a dead end if it is served by only one staircase. Above a certain height, generally three floors in residential and four in commercial buildings, it is necessary to provide alternative escape to a roof or by galleries to adjoining buildings. In a great number of cases, alternative means of escape upward to a roof or collection area may be practical as a last resort, as in the case of multi-storey buildings of high population where down-going staircases might become dangerous with congestion. However, it is not ideal and such an arrangement may cause confusion in the minds of users during a time of possible panic. Safety lies in familiarity with obviously placed and frequently used stairs. Figure 11 illustrates some common means of

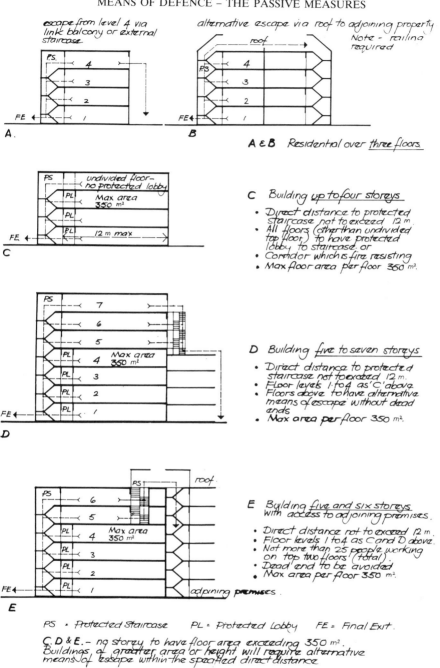

Fig. 11 Vertical escape, single staircase buildings – Greater London Council.

vertical escape and it should be noted that lobbies between floor areas and staircases form an important feature of safety where the choice of an alternative escape route does not exist. The design of such lobbies is discussed below.

Widths of escape routes

Having decided on the relevant principles and direction of escape, the designer must establish the widths of routes required. These depend on the number of users and the time to be allowed for evacuation. Places of assembly involving large unprotected spaces require simultaneous escape for all occupants to be completed in a minimum time. Places of work, particularly large commercial buildings, workshops and the like, might be better served by phased evacuation. It is important to consider carefully the size of an external space into which people will disgorge when large numbers are involved. Crowd panic can have disastrous results on staircases and around exit points as the recent history of fires in public places has so cruelly demonstrated.

The capacity of an escape route is determined by its narrowest width, usually at the doors, except in auditoria where they must open flush with the enclosing walls. Although the width is usually prescribed on the basis of occupancy and use, it can be calculated using the formula:

$$\frac{\text{Sum total of opening width in inches}}{22} + \frac{\text{Width of largest opening}}{} = \frac{\text{Number of occupants}}{100}$$

The total width of exits is calculated on the assumption that any one of them might be unusable. The formula is based on the premise that 100 people can pass through an opening 22 inches (now 530mm) wide in 2.5 minutes – the standard discharge rate for openings. It is accepted that a percentage increase in time should be permitted when considering staircases. The flow rates of people through escape routes can also be calculated using the following figures:

● Corridors – 1.5 people per metre width per second.

● On up-going stairs – 1.1 people per metre width per second.

● On down-going stairs – 1.15 people per metre witdth per second.

Ventilated lobbies

The function of the ventilated lobby separating staircases and escape routes from floor areas needs to be understood. It will be recalled from chapter 1 that the presence of fire or hot smoke in an enclosed space builds up pressure higher than the ambient, forcing smoke out of the fire zone into general circulation areas and, if preventative measures are not taken, beyond and into the main escape route. The escape route must be kept free from smoke. This will depend on the release of the thermal pressure and the smoke with it. Ventilated lobbies perform this function and provide an invaluable intermediate safety zone.

The diagrams in fig. 11 show the situations where ventilated lobbies are required. They are a necessary protection to staircases in all buildings where no alternative means of escape exists. In multi-storey buildings they are essential because of the extra risk of smoke emission into staircases and because of the large number of people who might otherwise face the danger of being trapped after blockage of the alternative route.

Exhaust of smoke from a ventilated lobby incidentally helps to indicate to the fire brigade the location of a fire, whilst the lobby itself will provide a safe base from which to tackle the seat of the fire. It must be remembered though, that hoses run through fire doors hold them open and prevent them acting as effective barriers to smoke. This is a significant point to bear in mind when locating wet or dry risers, which should ideally have their hose outlets as close as possible to the area at risk but in a protected zone.

Doors as smoke barriers

Because of the build up of thermal pressure, smoke will be forced round the edges of closed doors at such a rate as to render a corridor or lobby unusable within a period of between five and fifteen minutes. Every time a door fails as a barrier, smoke gains access to another reservoir, extending the chain of spaces through which the wave of fire will rush. The value of ventilation to reduce this pressure whenever possible cannot be overstated. Of value, too, is a well specified door in its role as a barrier to smoke, not to be confused with its function as a fire resisting door which concerns heat and flame resistance, which are discussed later in this chapter. Smoke may be cold, in which case preventing its passage is a matter of efficient selfclosing devices and effective edge seal, best achieved by high quality joinery and, if necessary, edge brushes. The use of intumescent strips is not a panacea in this respect, for many types require considerable heat in order to swell and do their work. Also, whilst some strips have a gap filling effect, others unfortunately clamp the door in a closed position and actually prevent escape.

Unfortunately, users often create risk by wedging doors open because they occur in positions inconvenient to the pattern of most common circulation: a problem too often experienced in main internal lobbies and corridors. The answer to this lies in anticipating the pattern and ensuring that doors occur at either end of, rather than in the middle of, the busiest routes.

Staircase enclosures

Staircase shafts must not be allowed to become reservoirs of smoke. Ventilation is a commitment. The common dog leg form of stairs acts as a constraint to the normal buoyant movement of smoke. Smoke, particularly if it is cooling rapidly, can then gather on landings and find its way into lobbies and passages of floors above the seat of the fire. For this reason ventilation is required not only at the top of the staircase shaft but at every floor level, again it is helpful to the fire brigade for locating the area in which the most immediate action is needed. Internal staircases in multi-storey buildings will require smoke exhaust ducts adjacent to them, with ventilators on each landing. In buildings of moderate height, natural ventilation via shunt ducts may be sufficient for the purpose but in tall buildings, loss of buoyancy as the smoke cools causes the smoke to become static and it is necessary to rely on artificial ventilation.

The hidden spaces

In chapter 1, sufficient was said about the spread of fire in confined spaces, such as ducts and ceiling voids, to give particular meaning to the need of their defence. These are the places which suffer casual design and neglect in building and maintenance, but where fire creeps insidiously or races furiously. Consider the list of worst offenders.

- Suspended floor spaces.
- Suspended ceiling voids in rooms and corridors.
- Main vertical service ducts.
- Gas meter rooms.
- Electrical intake and distribution spaces.
- Little-used store rooms.

All these have one problem in common: services that pass through them from area to area, placing at risk the integrity of separation. They all require occasional access for maintenance so that their protection is damaged. Ceiling voids suffer from the special disadvantage of normally lightweight construction exposed to the maximum of heat.

The regulations define the protection that must be afforded by the enclosure of such spaces, but a check list of points for the designer to bear in mind is helpful.

- No hidden spaces may provide communication between rooms, sections of corridors and lobbies, nor between floors or any other spaces which are themselves separated for reasons of fire safety.

- If such spaces contain services, these must be carried or sheathed in non-combustible materials, or the enclosure itself must be classified as fire protected.

- Access by door, hatch or panel must be fire protected; doors must be self-closing and labelled 'to be kept shut' where they occur in any circulation area.

- No such access should be provided from fire escape stairs.

- Services presenting a hazard, one to the other, should be in separate spaces – e.g. gas and electric services.

- Fire stops in ceiling voids must not be bypassed by combustible materials in the structure of the ceiling at its junction with fire protecting screens or partitions.

- Air handling trunking must not form a link between protected or separated spaces; automatic smoke dampers must be provided.

- Suspended ceiling materials must be resistant to spread of flame.

In the frequent haste and confusion of construction, the feeling that 'out of sight is out of mind' often prevails. The enquiry into a disastrous fire in a comparatively new Japanese hotel revealed that extensive spread of smoke and fire, directly resulting in the loss of many lives, was due to careless workmanship in the building of common ducts. The architect or building supervisor must take great care to ensure that the following items of work are thoroughly well executed:

- Concrete floors in vertical ducts completely sealed and made good in concrete to the full thickness of the slab around all services passing through it. How this is to be done should be considered at the design stage or the method discussed and agreed with the contractor on site.

- The material of the services provided is as specified: e.g., in unprotected ducts – drainage in cast iron, electrical wiring in steel conduit or trunking, etc.

- Block or brick walls enclosing ducts well built around penetrating services: no cracks must remain.

- The timber of framed duct enclosures completely sheathed in the correct thickness and quality of incombustible material: e.g., two layers of plasterboard properly lapped.

- No passage of air between vertical ducts or suspended ceiling voids.
- Ducts thoroughly cleaned on completion.

Spaces of special risk

In most buildings, large and small, there are spaces which, by their use, present a particular hazard or have special safety requirements. In extreme cases, such as storage of liquid petroleum gas (LPG), fuel oils, petrol, explosive and noxious or gaseous chemicals, they are subject to specific regulations and byelaws. Dangerous work and processes, too, are carried on in regulated conditions requiring structures, protection and means of escape stipulated by special Act of Parliament; the buildings are scrupulously inspected and the regulations rigorously enforced. Such situations fall beyond the scope of study in this book, but no architect or engineer should embark on the design of facilities which contain the least suspicion of such elements of risk without specialist knowledge or prior consultation with specialists. Within the more routine scope of a designer's work fall:

- Car parking.
- Plant rooms for boilers and air handling equipment.
- Communication and exchange rooms.
- Switchgear rooms.
- Transformer rooms.
- Standby generator plant.
- Lift shafts and motor rooms.
- Escalator machinery spaces.
- Water treatment and purification plant.
- Refrigeration and temperature control rooms.
- Incineration plant.

Where such accommodation is of ancillary status in a large building of some other use, rather than in a special detached structure of its own, it is usually banished to some remote and often inaccessible part – e.g. in basements, on roofs or in special service floors. If there is a disturbance factor as well as a hazard factor attached to its use, then the enclosing structure is almost automatically given by the designer the characteristics typical of structures of compartmentation – that is, weight, thickness and incombustibility. In so doing, the designer must not overlook other provisions of safety which belong to certain uses of space, for example:

• Natural or artifical smoke ventilation	Car parks, all main plant rooms below ground, minor plant and equipment rooms in central service cores.
• Fire fighting access	Free of encumbrance for personnel and equipment, especially important for parking areas and accommodation for heavy plant and machinery using oil, gas or electric power
• Means of escape other than the usual means of entry	For parking areas at any level and any plant accommodation in which the travel distance becomes a critical factor.
• Separation from public areas and internal circulation	Fire resisting doors and lobbies between parking areas and plant rooms that represent special risk.
• Separation of functions within plant rooms	Electrical switchgear, communications equipment, etc. isolated from gas- or oil-fired boilers and other combustion plant.
• Completion of compartmentation	By automatic fire doors and shutters in very large areas of fire hazard such as basement car parks and whole floor service zones (active measures).
• Fire prevention installations	Such as sprinkler, surveillance and warning systems. Almost any area of special hazard, especially basement car parks (active measures).

Protection of the fabric – the classifications which apply

In selecting the materials of a building, the designer will be concerned with two basic objectives in terms of fire safety. The first is the maintenance of structural safety; the second, resistance against spread of fire. To this end, specific aspects of a material's performance are relevant and in the United Kingdom they are tested, measured and classified in accordance with the terms of the all-embracing BS 476. These aspects are:

• Non-combustibility (BS 476 Part 4) – Designation which applies to materials which will not undergo thermal decomposition with the release of heat above a certain level or manifest continuous flame ignition for a period of more than ten seconds.

- Ignitability (BS 476 Part 5) – Classification of a combustible material as 'X', one that will ignite easily, or 'P', one that will not ignite easily when in contact with a small flame; whether it will merely scorch, burn slightly and go out or, on the other hand, flare up.

- Fire propagation (BS 476 Part 6) – Classification comparing the contribution of combustible building materials to the growth of fire, 'X' indicating high contribution, 'P' low.

- Surface spread of flame (BS 476 Part 7) – Classification in relation to wall and ceiling finishes indicating the tendency of materials to support the spread of flame across their surface, an occurrence that need not in fact affect the substrate or body of the material at the same rate as the face. Classification is given according to the rate and distance of spread of flame: best performance, Class 1; worst, Class 4. Class 0 is a term used in the Building Regulations for non-combustible materials which do not support any spread of flame (e.g. glass). Some typical spread of flame classifications are given in fig 12.

- Fire Resistance (BS 476 Part 8) – A time rating indicting the period for which an element of building construction retains aspects of its normal performance when subjected to the effect of standard fire. Failure may be of:

Stability of load bearing and non-load bearing elements.

Integrity, i.e. failure as a smoke barrier to hot gases and flame.

Insulation, i.e. failure by allowing the passage of sufficient heat to transfer fire.

The main elements of construction to which fire ratings will apply include walls and partitions (load bearing and non-load bearing), floors, flat roofs, beams, suspended ceilings, protection of steel frames, doors and shutter assemblies and glazed screens.

Protection of the structure

Fire resistance is one of the most important factors influencing the choice of the structural system to be used in any new building, large or small. Use and size determine the fire rating of a structure required under the terms of the regulations, as illustrated in fig. 3, and the achievement of that rating depends on various factors, for example:

Material	Class
Plasterboard	1
Plaster with 0.25 mm wallpaper	1
Woodwool slabs conforming to BS 1105: 1963	1
Timber, hardboard or fibre insulating board, good surface or impregnated	1
GRP (flame retardant)	1
Rigid UPVC sheet (when it can be classified)	1
Polycarbonate	1–2
PVC foam	1–3
GRP sheet (various types)	2–3
Hardwood, softwood or plywood, density greater than 400 kg/m^3, untreated	3
Hardwood, softwood or plywood, with oil-based or polymer paint	3
Wood particle board, untreated or with oil-based or polymer paint	3
Hardwood, untreated or with wallpaper or with oil-based or polymer paint	3
Polyurethane foam	4
Cellulosic fibre insulating board, untreated	4

Fig. 12 Some typical results of surface spread of flame tests.

- In situ and precast concrete:
  Dimensions of members (column size, wall thickness/length).
  Thickness of concrete cover to steel reinforcement.

- Structural steelwork:
  Thickness of in situ concrete casing.
  Fire resistance of insulation of enveloping material such as block-work, plasterboards and proprietary casings to columns.
  Fire resistance of suspended ceilings to floors in compound structures.
  Use of water-cooled hollow tube structures.
  Use of concrete-filled tubular structures in which the steel and concrete share the total responsibility for stability.

- Timber frames:
  Protective sheathings such as plasterboard or proprietary casings.

Fire retardant solutions and intumescent paints.

Sacrificial timber, i.e. extra thickness of timber surplus to structural requirements so as to allow for the formation of a layer of protective charcoal.

● Masonry:

Thickness.

Mechanical constraint over length and height (stability) provided by steel or concrete ties.

From this list it is evident that in many instances it is necessary to consider both the materials and method of assembly together and not just the material alone. This is particularly so in the case of floors and flat roofs, which are subject to special mention and testing in BS 476 Part 8, whilst Part 3 is devoted to the subject of external fire exposure roof tests.

Roofs and ceilings

As described earlier in this chapter and in chapter 1, roofs are exposed to ignition by heat from below or heat and embers from adjoining buildings, which are likely to lead to fire penetration to the inside. As far as possible neither the soffit nor the top surface of a flat roof should be faced in materials which significantly contribute to spread of flame already caused by the emission of volatile gases from the burning contents of the building.

In the case of flat roofs, and in some circumstances pitched roofs, ceiling materials will have to comply with limited spread of flame requirements as laid down in the Building Regulations. This is generally to Class 1, but in conditions of severe risk may be Class 0.

The use of suspended ceilings has come to play an important part in the protection of the structure of both roofs and floors against fire from below. But there is of course no protection from fire within a roof void. The contribution to the overall fire resistance of the floor or roof will depend upon whether the suspended ceiling is of a continuous (jointless) form or in panels, the insulation afforded by such panels, the spread of flame rating of their top surface, the nature and integrity of their fixings, whether or not the suspension grid or frame is exposed and the provision to prevent damaging thermal movement of the frame. Quite obviously the protective performance of a ceiling would be impaired by the insertion of flush lights and other fittings, unless they are specially designed to have the same fire resistance as the remainder of the ceiling. Any weakness in the assembly as a whole which might cause a small part to collapse will immediately expose the space above to the risk of flash-over and the further fall of burning debris on to the floor area below.

The risk of falling burning debris must also be taken into account when selecting the finish for the top face. It is preferable that the finishing coat of a roof has a low melting point so as to collapse or run off before it ignites.

Roof lights

The principle of smoke venting has been described in this chapter in relation to means of escape and the reduction in the risk of fire from smoke explosion. Some roof lights are intended to form vents when destroyed by heat. However they can in fact assist in the spread of flame if they are recessed above the soffit of a suspended ceiling, are closely spaced and fail to collapse. In such circumstances there is the risk of fire spreading from one to the other as gases burn in one recess setting off spread of flame across the ceiling surface to the next. Where recesses occur they should be deep enough to contain such burning, well spaced and laid out in a pattern which discourages this phenomenon. It should be remembered that very large roof lights which collapse will ventilate smoke around the perimeter of the opening but will draw hot air from the space below through the centre, thus setting up air circulation, feeding the seat of the fire with oxygen. Large roof lights then should not be designed to collapse but should have permanent or mechanically operated vents around the perimeter or, if they are steeply pitched in section, at their peak.

Doors

The function of doors as barriers to smoke in the protection of circulation spaces was also discussed earlier in this chapter and the point was made that this particular rule should not be obscured by the wider function of fire resisting doors – those which, together with their frames, are capable of resistance against collapse, flame penetration and excessive temperature rise for a given period. (BS 476 Part 8). The term 'fire check door' is commonly used and this can cause confusion because it is a generic description of any door intended for some degree of fire control and has no meaning within the terms of BS 476 Part 8. In fact all doors are capable of checking the progress of fire even for a period as short as six minutes, as is the case in a standard hollow core flush door.

Doors must be thought of together with their frames and iron-mongery and it is therefore sensible to think in terms of 'door sets'. Doors are tested fully fitted with their ironmongery and under conditions of thermal pressure. This was not always the case and new testing standards were introduced in 1972 (i.e. Part 8 of BS 476). However, doors continue to be manufactured and marketed with fire ratings based upon the original standard tests drawn up in 1953. The dichotomy created by these dual standards gives force to the need for the designer to understand the significance of the details of fire resisting door sets.

The metalwork of ironmongery conducts heat and transfers fire. Therefore avoid heavyweight hinges, the leaves of which reach from face to face. Mortice locks and latches should be slim, well concealed and remove as little timber as possible. Door closers, which present the largest area of heat conducting material, should be surface fixed. It is often thought that closers are a prerequisite of fire resisting doors. This is not necessarily so and they are required primarily where there is no other means of holding a door in the closed position; such as in the case of double doors. Again it is thought that double leafed doors need rebated leading edges. However tests indicate that this is not so provided intumescent strips are inserted down the edge, but one must remember the problem that cold smoke will not activate intumescent strip into creating a seal. Brushes along the edges might be used instead.

Increasingly, ventilation grilles are required in doors, particularly to meet the recommendations for depressurisation on escape routes. Such grilles can be treated with intumescence to act when the temperature builds up to a dangerous level. The introduction of intumescent materials has led to changes in attitudes towards door stop and rebate sizes and the fixing of glass panels in doors. The use of heavy stops and beads frequently leads to clumsy workmanship; small beads bedded in intumescent paint and intumescent strips rebated into frames lead to neater and less vulnerable details.

Glazed panels in doors and side lights are of great help in observing fire conditions in addition to the safety they provide in normal use. The subject of glazing is discussed later.

If a designer is able to keep in mind during the design stage, the ultimate fire protection rule of the doors he is positioning, the irritations of inappropriate ironmongery assemblies and clumsy details on doors of importance might be avoided. Remember then that door sets must be considered in the context of their use and location, broadly as follows:

● Doors in regular use but closed between times.

● Doors in constant use which need to be kept open until a fire occurs.

● Doors which are only used for very special occasions such as access or maintenance (ducts, etc.).

● Doors used for escape only which might otherwise present a security problem.

- Doors located on escape routes and having the dual purpose of smoke containment and fire protection.

- Doors which when closed form part of the compartment wall or enclosure with which they need to provide an equal fire resistance.

In all circumstances, a rule worthy of adoption is that a door be closed when fire occurs.

Glazing

The internal use of glazed screens to provide light, clear vision and a sense of space becomes a costly business when fire protected areas are involved. Georgian wired polished plate or rough cast glass is the usual material in such areas. Differential thermal movement between the glass and wire will cause the glass to crack when subjected to heat but the wire holds the cracked parts in place, so the sheet as a whole will restrict the movement of heat and smoke so long as the frame holds up. But it provides a solution which is often unwelcome to the designer in such circumstances, particularly when the size of each sheet must be strictly limited and where heavy timber sections are required to meet the requirements of the regulations. Similar problems arise on external elevations, where internal escape routes and external fire escapes stairs have to be protected.

Recently, new types of heat resistant glass have been introduced. One of these is a lamination of glass and clear intumescent sheet. Another is a thick sheet (6 to 7 mm) of special glass with a low thermal expansion coefficient and a high softening point, which together give an enhanced thermal and mechanical strength resulting in fire resistance of up to three times that of normal glass. The degree of fire resistance achieved under the tests laid down in BS 476 Part 8 will of course depend upon the type of frame and the size of the panel, which may be larger than can be achieved with Georgian wired glass.

Timber

Despite the fact that timber and timber-based products cannot be rendered non-combustible, partial protection can be achieved by

- Pressure impregnation in workshop conditions with water-soluble inorganic salts.

- Surface treatment on site with intumescent paints or other coatings which form a protective glaze or produce vapours that interfere with combustion chemistry.

These treatments may be used to upgrade normal spread of flame ratings from Class 3 or 4 to Class 1 (BS 476 Part 8) and in some instances when used in combination, Class 0, and may possibly achieve one hour fire resistance. This brings them within the requirements of Part E of the Building Regulations.

Impregnation is the more permanent of the treatments, although there is some risk of leaking salts in humid conditions and corrosion of metal fixings (which could require anti-corrosive treatment). There will be some limitation on the size of members that can be treated and timber sections should be milled first so as to avoid cutting away the best protected outer layers of timber. Although the treatment improves resistance to fungal and insect attack, there can be loss of strength particularly in reaction to sudden impact, accelerated under conditions of heat. Timber products tend to be resistant to impregnation and the adhesive content of boards can be adversely affected. Surface treatment has the maximum effect on surface flaming but the minimal effect on the main bulk of the timber. However, brush application on site seals joints and protects fixings; a valuable attribute. Many of these treatments are decorative and make a suitable finish to timber already impregnated with solvents. Intumescent applications are sensitive to moisture and become ineffective in humid conditions such as swimming pools, kitchens and laundries.

Communications and safety

Communication for fire safety is not merely a matter of symbols, signs, lights and bells. It embraces the entire gamut of comprehensive design so that people can be confident about assuming the obvious – and can be correct, with instant recognition, instant understanding and, thus, safety.

On the broad level of total building layout and design, instant comprehension is of paramount importance. Tortuous circulation routes, unexpected barriers, concealed lifts and stairs and contrived ad hocary only serve to confuse building users in normal circumstances. At times of crisis they cause suffering and death.

The purpose of good communications should be to inform upon five important points:

- The possibility of fire.

- The means of preventing fire.

- The advent of fire.

- The routes of escape from fire.

- The means of fighting fire.

For each of these points there are recognised means of communication, audio and visual. Most of them, however, fail to make an instant impact or to be entirely comprehensive in times of stress. Audio warning systems break down or give false alarms and are treated with casual disbelief. Sign systems are generally ill-designed, unfamiliar because of inconsistent use, unnecessarily wordy, unclear (perhaps even in a foreign language) and, when most needed, often obscured by smoke.

The effectiveness of alarm systems, covered in chapter 4, is largely a matter of management and maintenance. The success of sign posting is very much the responsibility of the designer and the fire prevention officers together, but the latter will be restricted by local practice and interpretation of national codes which naturally assume that at a glance everyone reads English. It is fortunate that the standard range of signs of mainly European inspiration is gradually becoming accepted in this country. To the designer, for whom written notices in huge letters (usually fixed in the middle of or over doors) are an anathema, these symbols will be a great advantage. In selecting and placing signs:

- Use internationally accepted symbols instead of typeface notices wherever possible.

- Keep signs low enough to avoid smoke obscuration.

- Avoid signs on doors or on flanking walls; open doors will put such signs out of sight.

- Locate signs so that they are illuminated by secondary lighting if not self illuminated.

- Fix signs so that they are not easily removed or defaced.

If the need for signs is anticipated, their proper integration with the design becomes a possibility – preferable to the application of signs to the finished decoration at the belated instruction of the fire officer. A check list based on the purposes of communication given above will assist.

The possibility of fire
Identify and signpost the possible cause or source of fire:

- Flammable materials stored – fuels, chemicals, gases, etc.
- Materials which become flammable under certain conditions of storage – grain, industrial waste, various fibrous products.
- Processes which produce fire risk – chemical processes, many types of fabrication work, welding, etc., milling, (production of volatile dusts), grinding (sparks).

- Accommodation with particular risk – electrical intake and meter rooms.

- Incidental risks – display and stage lighting, audio and visual display equipment, cooking, etc.

- Building services – machinery, heating equipment, etc.

The means of preventing fire

Signpost instructions on safety precautions and preventative maintenance:

- Preservation of fire integrity – maintenance of clear escape routes, closure of fire doors, no smoking areas, restriction of parking, prevention of dumping.

- Location of preventative installations – electrical earthing, lightning conductors, switchgear, circuit breakers, gas valves, stopcocks, sprinkler controls, smoke exhaust controls.

- Instructions on maintenance – checking building services and moving plant, sprinkler operation, extractor fans and all fire fighting equipment and alarm systems, fire doors and shutters, lifts and escalators, emergency lighting, stand-by generators.

- Notices of guidance – handling of materials, operation of machinery, methods of storing and stacking, preservation of ventilation.

The advent of fire

Signal the occurrence and location of fire:

- Local audio and visual alarm signals – coupled to automatic detection equipment, sprinklers, etc., and to manually operated control station.

- Surveillance by close circuit TV – to local monitoring and control station.

- Surveillance through co-ordinated local response systems, CCTV, etc. – controlled by remote monitoring station with direct lines to fire brigade.

The means of escape from fire

Instruct and signpost so as to provide background knowledge and immediate information on optimum or alternative routes and on preservation of life:

- Direction signs on primary routes – signposting to nearest safety zone, staircase and final exit, all with simple symbols.

- Notice of assembly points – for zonal evacuation.

- Notices of advance information – in special places of work, public accommodation, hotel rooms, etc., giving information on alarm signals, action to take, direction of escape routes, nearest staircases, etc.

- Notices of location of special escape provisions – chutes, folding ladders, roof walks, helicopter lift-off, etc.

- Notices locating and instructions on use of protective equipment and clothing – breathing apparatus, fire suits, fire blankets, first aid supplies, stretchers.

Means of fighting fire

Communicate information on every aspect of fire fighting procedures, equipment and control:

- Location signposts and symbols of equipment – hand appliances and their appropriate use, dry and wet risers, hydrants, hose reels, fire blankets, mobile equipment.

- Instructions on correct use of equipment – correct use of hand extinguishers for particular types of fire.

- Use of lifts – fireman's lift.

- Isolation of services – action on gas and electrical supplies, manual over-ride on automatic control of plant, ventilation, etc.

- Isolation of seat of fire – closure of doors and windows, removal of furniture, goods and materials which might contribute to fire growth.

- Direction and control of fire fighting operations – loud hailers and intercom for instructions to building occupants and outside crowd and traffic control.

- Communication with public services – fire brigade, police, public utilities and hospitals.

'. . . low-flicker frequency. . .'

Chapter 4

MEANS OF DEFENCE –
THE ACTIVE MEASURES

When a fire occurs there ought to follow a chain of events of such necessity as to justify description as 'the law of active measures'.

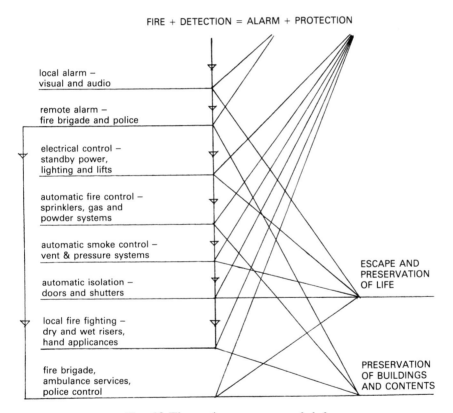

FIRE + DETECTION = ALARM + PROTECTION

local alarm –
visual and audio

remote alarm –
fire brigade and police

electrical control –
standby power,
lighting and lifts

automatic fire control –
sprinklers, gas and
powder systems

automatic smoke control –
vent & pressure systems

automatic isolation –
doors and shutters

local fire fighting –
dry and wet risers,
hand applicances

fire brigade,
ambulance services,
police control

ESCAPE AND
PRESERVATION
OF LIFE

PRESERVATION
OF BUILDINGS
AND CONTENTS

Fig. 13 The active measures of defence.

From the moment of fire detection, many of the activities listed in fig. 13, linked by a common control system, occur simultaneously. Alternatively, with certain hazards or particular conditions of escape, reaction to fire detection may be programmed to occur in a particular sequence, be phased or have in-built delays.

Detection and alarm

Depending on the building's use, size, value of contents or number of occupants, detection may be represented by simple local response equipment, as in domestic premises, or by a total surveillance system designed to meet the particular needs of a large building or complex such as a major hotel, factory or warehousing facility. Such systems might incorporate a wide range of detection equipment, each type selected for its suitability to operate in the local conditions and each one signalling to a central control or monitoring point, from which the appropriate warning and protective measures will be activated. Signalling can be extended via telephone lines to an off-the-premises monitoring station run under a full-time surveillance contract or, in special circumstances, direct to the fire brigade.

Many systems operated under contract include surveillance for protection against intruder, whilst the most sophisticated of in-house operated systems can include monitoring of environmental conditions, the performance of heating and air conditioning plant, the operation and location of lifts, the integrity of security zones and even the movement of authorised personnel. However, fascinating as the complexity and ingenuity of such all-embracing surveillance systems are, we are concerned here only with those parts which detect and signal the advent of fire.

Systems should be designed by firms specialising in this work and must conform to BS Code of Practice 5839, and, where applicable, the Rules of the F.O.C. The fire brigade and insurers should be consulted. It should be noted that systems which are solely automatic are not acceptable under the terms of the Factories Act 1961 or the Health and Safety at Work (etc.) Act 1974 and that manual operation should be possible.

Types of detectors

A variety of detection mechanisms are used which involve the completion of an electric circuit as a result of the thermal expansion of metal, liquid or gas or as a result of the break in a fusible link. Other types involve the change in the electrical resistance of conductors, or the change in voltage produced between two thermocouples.

Heat detectors
● Fixed temperature – Designed to operate at a pre-selected temperature

● Rate of rise/fixed temperature detectors – which operate when an abnormal rise of temperature occurs or when a pre-selected level is reached

Smoke detectors
- Ionisation detector – A chamber containing electrodes, connected to a supply. A small radioactive source ionises air in the chamber, turning it into electrically charged particles (ions) and so allowing an electric current to flow. Smoke particles attach to some of the ions and so reduce the current, activating the alarm.

- Optical or thermo-electric detectors – A light beam passes close to or falls on a photo-electric cell. Smoke entering the detector scatters the beam, changing the amount of light which falls on the cell and setting off the alarm.

Combined heat and smoke detectors
- Infra-red detector – Consisting of an emitter transmitting a wide angled beam of infra-red light which falls on a remote photo sensor. Loss of strength in the beam caused by smoke or thermal turbulence registers a fire signal upon exceeding a pre-determined level. Correction for temperature or accidental blockage can be built in.

- Laser beam detector – Consists of a projector sending out a narrow laser beam which is reflected on to a photo-cell receiver, partially masked so as to contain the incident spot of light in a 'no fire' position. The alarm is activated when the beam is deflected on and off the spot by the slightest turbulence from heat or smoke particles, but it responds only to the frequencies which would be given by a fire.

Flame detectors
- Infra-red, scanning or defined area detector – A photo-electric cell, the frequency of which is fixed to respond to the low-flicker infra-red radiation characteristic of a diffusion flame. Scanning detectors rest in position when a signal is picked up and if the radiation flicker pattern persists, then the alarm is given, if not, the scanner moves on.

- Ultra-violet detector – As for the infra-red detector, this detects the ultra-violet light emitted from flames.

The choice of detector will depend upon circumstances and, very often, a number of different types will be required in combination to deal with the varying conditions found in one building or even within a single space.

Heat detectors are generally used in confined or limited spaces or where the seat of the fire risk might be close, where occupants might be smoking, cooking or using any fume-producing equipment. Activation levels must be set to obviate the effect of heating equipment or sunshine. There are different grades of unit to be specified to suit the height of the space involved.

- Grade 1: up to 9 m
- Grade 2: up to 7.5 m
- Grade 3: up to 6 m

Smoke detectors should normally be sited at the highest point in a space and are suitable for most situations unless smoke or fumes are generated by the activities carried on in the area. These detectors give early warning of partial combustion and signal the danger of fire before flame ignition occurs. They are useful in concealed spaces, such as ceiling voids, where smoke may gather even if its temperature has begun to fall. Ionisation detectors are more sensitive than optical detectors, responding rapidly to invisible combustion products in the early stages of a fire but responding slowly in dense smoke. They prove over-sensitive to fumes, dust fibres and steam produced by normal processes or vehicle engines. They are also affected by fast-moving air currents.

Flame detectors have special application in outdoor situations and installations such as chemical plant and stores containing liquid flammables, in which flame would spread dramatically and lead to explosion.

There is less experience of the combined heat and smoke detectors. The narrow reflected beam of the laser beam detector makes it suitable for industrial installations with difficult access, i.e. tall or long spaces such as cable tunnels. Infra-red detectors, on the other hand, with their wide beams, have special value where broad coverage of large internal spaces is required with the greatest economy. Small obstructions, provided they are non-moving, are of little significance. The danger of broad cover from a single unit lies in the risk of its failure, leaving a large area unprotected. This is less likely to occur with a number of normally placed smoke detectors. Nevertheless, it is important that in a comprehensive surveillance system any equipment malfunction is immediately signalled.

Alarm

In large automatic systems, detectors will be grouped in zones so that the location of the fire can be identified. For the use of personnel in or escaping from the area of an otherwise undetected fire, it is essential that manually operated alarms are also located in the same zones and linked with the automatic detectors to the indicator panel.

The alarm itself might signal immediate total evacuation, as in places of public assembly, or it might be phased to give a general standby warning (indicated by an intermittent call sign) together with an immediate evacuation instruction (a continuous call sign) from the area of first danger. Such systems minimise the rush to escape routes and give time for proper close down of machinery and organized grouping of personnel at their allotted fire stations. Indeed in some establishments it would be ill-advised for occupants to leave their work place at all unless the danger of fire reached a degree of imminence. Where the system is linked to a remote monitoring station or direct to the fire brigade the alarm would be activated at that location and, again, the area of the fire identified.

The alarm signal must be one which can be readily identified by occupants, especially against a background of other intrusive noise such as machinery. Visual alarm signals should show concurrently with audio alarms, particularly where working conditions call for the use of ear muffs. The deaf must not be forgotten.

Electrical supply

Of the automatic protection which is immediately activated by the detection system, mention must first be made of the standby generators and emergency power system necessary for any building of high security or special function, such as hospitals, factories and laboratories, and without which the other automatic defence systems may fail to work. There is a moment's pause between the cut off of normal supplies and cut in of standby generators. Such a cut in the supply will affect other plant and machinery, temporarily stopping lifts, air conditioning and exhaust systems, cutting out electric lighting, releasing magnetically restrained fire doors and so on; all likely to cause confusion and anxiety. It is important then that proper phasing is built in to the electrical control with manual over-ride switch gear for operation by appointed staff or the fire brigade. The manual over-ride must be located in an obvious position and easily accessible. It must be possible for parts or the whole of the building to be isolated from all electrical supply if there is any risk of this contributing to the danger. It is possible that the cause of the fire may be in fact electrical or that fire might cause fusing, overloading and overheating, sparking in the presence of explosive gases and hence adding to an existing fire.

Escape and emergency lighting

Emergency lighting is required to provide a minimum level of illumination in the event of failure of the normal lighting. The standards are laid down in the Code of Practice BS 5266 'Emergency Lighting of Premises'. Escape lighting is that part of an emergency system which ensures that the means of escape can be identified and effectively used.

There are two basic types of emergency lighting:

- Slave luminaires or signs – These are operated from a central power system of batteries, charging device, generator or invertor and master switches. Fittings are cheap but installation costs high; separate wiring must be provided and this must be heat resisting. Failure of a subcircuit can cause the whole system to fail or the whole system to come on when only part is required. A relay system can overcome this problem but in itself is liable to failure.

- Self-contained luminaires or signs – A system in which the fittings only require connection to the normal permanent supply and which themselves contain all the necessary controls. Operating time in emergency conditions is for a minimum of three hours and highly efficient 4 watt fluorescent tubes are used, powered by rechargeable nickel cadmium cells maintained at full charge by a mains powered solid state module. A hold-off device is incorporated so that the lamp is unlit until the main power supply is interrupted, whereupon illumination is automatically provided. The batteries are recharged within 24 hours upon recovery of the normal power supply.

Fittings which operate only when the normal lighting fails, as described above, are defined as 'non-maintained emergency lights'. However, controls may be included which enable the lamp to operate at all times without affecting the charging of the batteries – 'maintained emergency lighting'. These would be used for example where permanent illumination of exit signs is necessary.

A common situation is that in which escape or emergency lighting is required in areas where switch lighting is normally provided. In this case a 'sustained luminaire' may be used which contains two lamps, one powered by the main supply, the second by an emergency supply.

The lowest level of illumination at the end of the emergency supply period is 0.2 lux which suggests that the minimum design level should be 0.5 lux. In critical areas such as assembly points, escape doors, fire appliance points, etc., forty times this level may be required. Along escape routes, the frequency of lighting must be sufficient to achieve the minimum level of illumination between each fitting, with special emphasis lighting provided at danger points such as bends in corridors, doors and staircases.

It is most important to remember that in smoke-filled spaces, luminaires may be used as guides to follow the escape route. It is therefore essential that each luminaire should be visible to a person standing under the one next to it. There is no point in placing fittings at a height at which they might be concealed by smoke and ideally each luminaire should offer some directive as to the position of the next. Exit and emergency exit signs (which should be illuminated at all times) should be between 2.0 and 2.5 m above floor level and positioned immediately above or adjacent to the exit door. Where this is not in direct sight line then subsidiary signs should be used to lead people in the right direction.

Lifts – the fireman's switch

Lifts in a multi-storey building on fire can be a mixed blessing. Used indiscriminately with collected calls taking precedence over car control, and thermal switches at landings being activated by fire heat, they can become death traps to those who see them as the quickest means of escape. Isolated from remote control and operated by trained fire fighters, the story is different. At least one and, in large buildings with banks of lifts, possibly two or three will be required to have fire service control. Those without should have automatic control taking them to the exit level the moment the evacuation alarm is sounded or standby power cuts in.

The 'fireman's switch' is a glass fronted box close to the lift under its control (preferably higher than normal control level) and located on the main entrance level or wherever recommended by the fire brigade. When activated it will isolate the lift from all landing and lift car controls and cancel all previously collected calls. The car will travel to the fireman's switch floor and stay there until required. From this moment on the lift car can be controlled only from within the car and without any form of call collection. Taken to any other floor it will stay there, doors open, until redirected by the authorised operator. An added factor of safety would be an emergency telephone in the lift with direct link to the control floor; this is especially valuable in high buildings.

All lifts should have emergency lights.

Sprinkler systems

An automatic sprinkler system is not merely a means of pouring water on a fire. It combines detection, warning and restriction. Essentially, it is a first-aid treatment, spotting the fire, spraying water on and around it, cooling the surrounding fabric and possibly extinguishing the blaze. At the same time it will sound the alarm and indicate the zone in which the fire has occurred. The exact type of system required will depend upon:

● The nature of the contents being protected and whether or not they might be seriously harmed by leak or accidental operation.

● The degree of hazard generated by the contents and the rate of water discharge required to make the fire area safe.

The latter of these conditions will relate to the available water supply, which must be checked at the outset, the building area and height and consequently the density and number of layers of sprinklers necessary to do the job. Although the use of sprinklers will affect the compartmentation of a building, neither the Building Regulations nor the fire brigade define any performance standards so that control officers have to rely on the F.O.C. rules and implementation to the standards accepted by the insurer's inspectors. These aspects of design are covered in detail in chater 5 which is concerned with insurance and the rules of the F.O.C.

There are various systems available to deal with the conditions which might be encountered.

The wet system

Figure 14 illustrates the general arrangement of the typical wet installation. It consists of a water supply from the mains or under pressure from tanks to a main control valve followed by an alarm valve. From this, a vertical riser feeds distribution pipes at each level and range pipes on which the sprinkler heads are mounted. The sprinkler head is a water outlet kept closed by a valve held by a thermally sensitive bulb or strut. When heat breaks the bulb or strut the valve is released and water discharged with force at a profiled disk which scatters the water in an umbrella pattern. The release of water pressure causes the alarm valve to open so that the water flows from the supply source. As this happens, water is introduced to a pipe from the alarm valve, creating in it a pressure which activates the alarm.

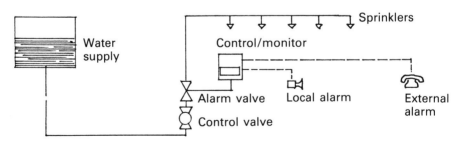

Fig. 14 General arrangement of wet sprinkler systems.

The wet system should not be used in areas where there is risk of freezing or where there is risk of damage to electronic or chemical installations.

The alternate system

This is one in which there may be water or air under high pressure in the pipework, selection of which, depending upon the conditions and contents being protected, is affected by a special shuttle valve at the main control. The danger in this system, when filled with air, is a delay of up to three minutes whilst the air is exhausted from the pipework before water will arrive at the sprinkler head. If the pipe work is already filled with water, the system will of course perform in the normal manner, the heads responding locally to heat or flame.

The pre-action system

This is similar to the alternate system but the problem of delay when the pipework is air filled is overcome by independent heat detectors which activate the shuttle valve and allow the entry of water into the system in advance of the normal thermal response of the sprinkler heads. The air in the pipework is maintained at a pressure considerably lower than that of the water supply which, when released by the advanced signal, compresses the air and reduces the dry volume in the pipework.

The cycling system

This a refinement of the wet or dry pre-action systems in which a network of heat detectors activate solenoid operated valves which turn the water supply on and off. If the sprinklers in an area suppress the fire, they will be turned off; if it should re-light, they will re-start.

The on-off sprinkler system

Individual sprinkler heads can themselves turn on or off, achieving a similar effect to that of the cycling system.

The deluge system

In principle this is similar to the standard sprinkler system, but the objective is to discharge large quantities of water over the entire protected area (rather than in a phased manner) immediately upon detection of fire. Rapid detection may be necessary, in which case the sensitivity of the sprinkler heads will be adjusted or heat or smoke detectors used, coupled direct to a shuttle valve control as on an alternate type system. Deluge systems are sometimes arranged to discharge solutions which foam at the sprinkler head. An interesting variation is the use of low level foam projection nozzles to cover floor areas in such places as aircraft hangars or other spaces where dangerous areas are screened by heavy equipment. Medium velocity sprays are used on deluge systems to cool fuel tanks or other explosive stores subject to danger from the heat of fires in adjoining areas.

High velocity (or water fog) systems

Essentially these are wet systems under high pressure, designed to extinguish oil or flammable liquid fires. Special sprinkler heads discharge a conical spray in which fine droplets of water, travelling at high velocity, emulsify and cool the surface of the flaming liquid, bringing it below the combustion temperature level.

Innumerable variations on these systems can be introduced to deal with the special requirements of a building or its contents. Even though the design of sprinklers is a highly specialised matter, knowledge and forethought on the designer's part will ensure that the general or special conditions for their installation are catered for at an early stage of the design process. To this extent a check list of points to remember is useful:

- The adequacy of the water supply must be checked.

- Standby generators or batteries may be required to maintain the operation of controls and pumps.

- In any one building, more than one type of system may be required, e.g. offices in special warehouses.

- Sprinklers will be required
  over escape doors,
  over escape and other stairs,
  in lift shafts,
  in stores and some plant rooms,
  possibly in ceiling spaces and ducts.

- Sprinklers should not be installed in electric intake rooms.

- Sprinklers may be required at several levels inside high storage racking or within special machinery.

- To avoid unnecessary damage, the water from sprinklers will need to be disposed of either with scuppers to the exterior or by a system of floor drains.

- Points for drain-off at low level and pressure testing at high level will be required.

- In large installations the main distribution pipes may be of surprising size and require careful co-ordination with other services.

- Thermal or structural movement joints may be required in main pipe runs.

- Access to all parts for regular inspection and maintenance is a necessity.

A final warning to designers is that the operation of sprinklers can cool a developed fire to a state of partial combustion so that heat operated ventilation systems fail to come into action and dangerous smoke logging occurs. Conversely the prior operation of an automatic smoke exhaust system can increase the circulation of air and intensify the heat of a fire, making it more difficult for the sprinklers to control the situation.

Foam injection

The use of foams is generally associated with heavy mobile fire fighting equipment or small hand appliances. In fact, foaming agent injection equipment can be fitted to water systems such as sprinklers, or, more commonly, wet riser branch pipes and hand hose outlets. Without doubt, foam is the most rapid and effective medium for fighting fires involving flammable liquids and toxic chemicals; in fact so rapid is the effect of complete saturation in foam that the expression 'knock down' is used to describe the extinguishing of a fire. The success of foam depends upon its low viscosity and stability, enabling it to form an insulating blanket which will smother fire, allow time to cool and prevent 'burn back' (i.e. re-ignition) and suppress the release of noxious combustion products. This particular quality makes foam a useful agent in preventing the discharge of dangerous vapour from the spillage of non-flammable toxic liquids. Another advantage is that foaming agents can be used with fresh or salt water: the reason for their wide use in fighting marine and especially oil rig fires.

Most foam liquids are protein-based, non-toxic and biodegradable. The exact composition will depend upon the quality required of the foam in dealing with particular circumstances; whether the emphasis is required on rapid knock down, resistance to burn back or suppression of vapour release.

Concentrations of agent will be between 3 and 6% so that the volume of agent stored will impose a limit on the quantity of foam which can be produced from one source and the time available for its application. It is for this reason that large mobile storage/fire fighting appliances are used, kept in a state of readiness close to a likely source of fire – such as airport facilities, large fuel and chemical storage plant or fuel handling equipment.Built-in foaming plant can be installed for the permanent protection of stationary hazards such as floating top fuel tanks and fuelling bays, these requiring regular and careful servicing.

On a smaller scale, hand appliances have application in cooking areas, garages and certain types of plant rooms.

Gas and power systems

Actuated by the same methods as sprinkler systems (smoke and heat detectors, etc.), built-in gas and powder systems also combine alarm and defence. They are used in situations where water spray or deluge systems might not be effective, could be dangerous or would cause damage to equipment and materials. Typical of such circumstances are processes or plant involving water reactive chemicals, spaces for delicate electronic equipment (computer rooms, etc.) and machinery spaces for the manufacture of delicate products. Even within the range of gas and powder systems, however, the use of one or the other could prove harmful. For example, carbon dioxide can cause freezing up of metallic parts in delicate equipment; powder can clog moving parts, conceal damage and be difficult to clean.

Gaseous and powder agents are of two types:

- Inerting agents – Suppress fire by the physical means of displacing the air in the area of combustion so as to cut off the oxygen supply and by reducing the temperature below the combustion level.

- Inhibiting agents – Combine inertion with chemical suppression, so requiring a less concentrated application than inerting agents.

Carbon dioxide systems (inerting)

Carbon dioxide, a colourless and odourless gas, is inert, non-corrosive and electrically non-conductive. Stored under high pressure at normal temperature or at low pressure and refrigerated, its vapourisation pressure is one and a half times that of air, so that it will readily discharge and disperse throughout the space to be protected, cooling as it expands if it has been stored under pressure. Discharge nozzles may be arranged for local application, i.e. in the immediate vicinity of a potential fire hazard, or they may be laid out evenly over a wide area for total flooding of an enclosed space. In many instances it is appropriate to use a combined system in which discharge is in the area of the greatest risk so that the gas cools and inerts the fire prior to spreading throughout the entire enclosure. Where there is air conditioning which might exhaust the gas and diminish its suppressive capabilities, a system of phased or repeated discharge can be introduced.

There are risks associated with the use of carbon dioxide:

- Occupants – Whilst a 30% concentration of carbon dioxide in the atmosphere is required to suppress most fires, a concentration of only 10% can cause occupants loss of consciousness within a couple of minutes, followed quickly by death. There should therefore be an in-built delay between the sounding of the alarm and release of carbon dioxide, sufficient time to allow escape.

- Static – The release of carbon dioxide can cause a discharge of static electricity from the nozzles to earth, possibly causing an explosion of volatile combustion products in the atmosphere. Where there is such risk, steps must be taken to earth the discharge system.

- Damage by pressure – Carbon dioxide released under high pressure into an unventilated enclosure can cause damage to the building fabric. Carbon dioxide is heavier than air and therefore best released at low level, for example from outlets in a suspended floor, near the seat of the fire. Ventilation should be at high level.

- Temperature drop – Sensitive electronic equipment may be damaged by a dramatic drop in temperature, as is caused by the release of carbon dioxide. Insulation or shielding of such equipment is necessary.

The use of carbon dioxide is most effective against hydrocarbon fuel and electric fires, in enclosed spaces that may safely be filled with gas or in small concentrations in restricted work spaces. Typical applications will include:

- Engine room spaces involving petrol or oil fuel.
- Oil-filled switchgear, transformers, alternators, etc., in power stations and substations.
- Computer suites and control rooms housing electronic equipment.
- Accommodation and storage for valuable or delicate goods, works of art, etc., which would be vulnerable to water.

Halon systems (inhibiting)

There is a wide range of halogenated hydrocarbon agents or Halons as they are known, all with confusingly similar and lengthy names. They are therefore indentified in a manner established by the American Army – by numbers, in which each digit represents the number of carbon, fluorine, chlorine and bromine atoms in a molecule; for example, Halon 1211, commonly known as BCF (bromochlorodifluoromethane) and Halon 1301, BTM. These two Halons are in fact the two in most common use.

Halon is a dry colourless, odourless, non-corrosive and non-conductive gas; it is an inhibiting agent which separates the elements of fire, causing a breakdown of the combustion process. Halons are stored at normal temperature under pressure, as are gas or vapourising liquids, and protection is provided either by local application or total flooding, as in carbon dioxide installations. However, concentrations required to extinguish fire are generally much lower than is the case with carbon dioxide, being as little as 3 to 5% for a surface fire of combustible solids; 10% or more might become necessary for a more deeply seated fire. The actual concentration for which a system will be designed must depend upon the nature of the hazard. The objective will be to extinguish the fire before it becomes deeply seated, so as to keep to a minimum the amount of agent that must be released.

Occupants should not be exposed to a concentration of more than 4% of Halon 1211 for more than one minute, or 7% of Halon 1301 for more than five minutes. These concentrations represent a reasonably low level of toxicity, leaving a good supply of oxygen in the atmosphere and allowing useful time for escape during the initial stages of a fire. However, in addition to the normal noxious combustion products, toxic substances can be produced in the process of extinguishing a fire by the breakdown of Halon into acid gases. As the ideal low concentration of Halon would be of little value in a well ventilated space, release of the agent should be preceded by automatic closure of doors or ventilation systems; the very steps likely to increase any occupant's risk of exposure to toxics.

There are then certain rules to be observed:

- All systems must be custom designed to suit the individual application (hazard, location and occupancy).

- Flooding systems are designed to operate in enclosed spaces and all openings, air extract plant, etc., must be closed prior to release of the agent. Spaces must be sealed to prevent escape of the extinguishing agent or combustion products and subsequently well ventilated before re-entry.

- Halon gas is odourless (unless scented), so that release by an automatic fire alarm system must only be possible when an area is unoccupied or where a two stage warning system is installed.

- Automatic release systems must have manual override control with status indicators on a central control panel.

Where local application of Halon is required, storage will be in small wall or floor mounted cylinders with the minimum pipe run to a single nozzle outlet. Examples of this type of application include paint spray booths, engine test cells, laboratory cabinets, quench tanks, dip tanks and oil-filled transformers. Flooding applications are used where a hazard is within a fixed enclosure, which will ensure that an adequate extinguishing concentration of agent can be achieved and maintained for an appropriate period of time; for example cable voids, motor rooms, relay rooms, computer rooms, tape and record stores, art galleries and libraries.

Powder systems (inerting)

Fire fighting powders are finely ground chemicals, constituents of which depend upon the type of fire to be extinguished. Readers will be familiar with the fire type classification A, B, C or D appearing on hand appliances.

- Class A – Solid fuel fires.

- Class B – Flammable liquid fires.

- Class C – Combustible gas fires.

- Class D – Flammable metal fires.

Class A, B and C fires are all effectively extinguished by mono-ammonium phosphate based powders; sodium bicarbonate or potassium bicarbonate based powders are also used for class B and C fires. These powders are essentially inerting agents but they have little effect on combustion temperature. Once the dust has settled on the area of the fire, its extinguishing action is virtually completed. Smouldering combustion may persist even though the recurrence of flaming is unlikely. It is important, therefore, that in designing automatic discharge systems, the quantity of powder stored is wholly sufficient to extinguish any possible fire.

Powders used on flammable metals (aluminium for example) act by fusing to the metal surface and forming an encrustation which has an inerting effect. Powders such as PVC borax and boron-trioxide are used in a limited manner, discharged in small quantities directly on to local hazards such as special machinery or equipment casings.

Storage of powders is maintained under pressure at normal temperatures. The means of discharge ranges from local application by hand held appliances to multi-point centrally supplied systems.

Powders have two great advantages: effectiveness and the fact that, upon discharge, they are visible, unlike gases which, unbeknown to occupants, might be present in the atmosphere in harmful concentrations. This makes them particularly suitable for domestic hand appliances and generally useful in a wide range of situations where the residual powder is neither harmful nor inconvenient.

Powders should NOT be used on.

- Low voltage electrical installations, relays, switchgear, generators or motor equipment.

- Any form of telecommunications equipment, radio, TV or similar installations.

- Computers, micro-electronic equipment and associated hardware and software storage.

- Tightly packed solid combustibles where deep seated fires might occur, such as record files, furniture, fabrics, furs, etc.

Whilst this appears a formidable list of areas in which powder may not be used, those in which it is effective and suitable are far more numerous and only a few need be mentioned:

- Boilerhouses and engine rooms;

- Oil storage tanks and oil rig modules;

- Domestic and commercial kitchen ranges, hoods, ducts and general service areas;

- Industrial installations involving milling, pneumatic conveyors, dip tanks, quenching baths, etc.

Assisted smoke control

Earlier in the chapter, the description of sprinkler systems ended with a warning that the rapid exhaust of smoke could in fact inhibit the operation of sprinklers or vice versa. Provision made in the course of design to deal with natural ventilation is unlikely to cause any such problem, but the introduction of mechanical assistance can do so. The involvement of a specialist designer is then important.

Apart from permanent natural ventilation there are various mechanical options open to the designer, selection depending upon the problems created by the building design and the degree of risk.

Mechanically assisted ventilation systems

All such systems can be activated electrically by remote heat or smoke detectors or from a fire control panel, or they may be triggered off when heat breaks a fusible link.

- Mechanically opened natural vents – Invariably roof mounted, these are opened by the lever action of counterweights or by pneumatic or hydraulic action.

- Smoke extracting fans – Usually roof or high level wall mounted and in some circumstances associated with ducts of limited length, fans can be used for natural ventilation and cooling but have an automatic cut-in activated by the fire detection system. Extract fans are the simplest means of preventing smoke logging of large spaces where there is direct contact with the exterior. The air velocity must be sufficient to overcome external wind pressures, but the air circulation so created can stimulate fire.

- Ducting extract systems – Most likely to be part of a building's mechanical ventilation system. Whilst of great value in evacuating smoke from special areas and spaces deep inside a building, danger lies in the possibility of spreading smoke via the duct system to unaffected parts of the building or of a fire in the ducts themselves. Air pressures at discharge points are unlikely to be sufficient to overcome adverse wind conditions. Good maintenance, always difficult, is a prerequisite.

Pressurisation systems

Although it is twenty years since pressurisation systems were first conceived and tried, experience remains limited, primarily because the circumstances in which they are appropriate are rather limited and generally precedential. However, research has led to the publication of the Code of Practice BS 5588: Part 4, 1978 which is concerned with the protection of escape routes from smoke. It does not cover shopping malls and similar public spaces.

The basic concept is similar to that used in hospitals and diagnostic units, where protection from dangerous infection is required, or in research laboratories where cross contamination between work stations must be prevented. The air in the space to be protected is held under pressure and exhausted either direct to the exterior or via spaces of lower protection priority and thus lower pressure. This graduation of pressure creates air movement which carries smoke with it along the predetermined routes, away from the protected area. Application of the principle depends upon proper selection of the priorities and provision of the correct atmospheric pressure. The pressure is measured in 'Pascals' (where $1 \text{ Pa} = 1 \text{ N/m}^2$) and must be calculated so that it is sufficient to keep the air movement going in the right direction, taking into account one or more open doors in the protected area.

The most important aspect of the protection of escape routes is to prevent smoke from entering staircases. To achieve this the air flow must be away from the staircases towards the perimeter of the building or an alternative ducted form of exhaust. Figure 15 illustrates typical patterns of air flow and the pressures that might be required to achieve these. The Code of Practice calls for an air velocity of 0.7 m per sec through an open staircase doorway. This requires a pressure of 50 Pa in the staircase. With the door closed, pressure could rise to 60 Pa, so making it difficult to push the door open: a problem overcome by the introduction of pressure release vents.

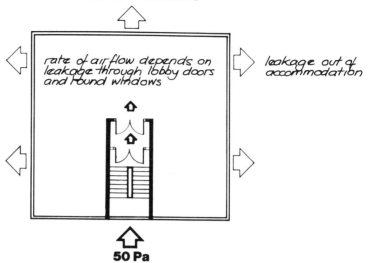

A - *air input from staircase via protected lobby*

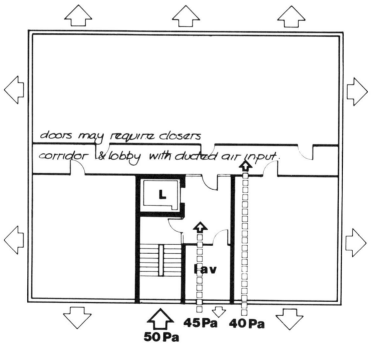

B – *air input from stair with reduced pressure input from lobby & corridor to compensate for leakage through lift shaft and lavatory*

Fig. 15 Smoke control by pressurisation.

To prevent smoke-laden air travelling via the staircase to other floors, the points of air supply should be distributed throughout the height of the staircase enclosure at intervals not exceeding three storeys (although there is successful experience of a single, larger fan supply at the head of a staircase) whilst exhaust should be from the building perimeter at every level. It is generally found that exhaust can be achieved by natural leakage through the external walls, but the provision of fixed ventilation is a safeguard. Fans supplying the air under pressure may, with advantage, be kept running at a low velocity the whole time as a contribution to the building's ventilation system. The velocity is then increased in response to control by the smoke detection system. This practice is of great assistance to maintenance as it creates awareness of the equipment's satisfactory function.

Designers of large buildings will undoubtedly become more involved in this form of smoke control and protection, but it must be remembered that this is a field of work requiring the services of specialist consultants. However, guidelines for the designer to bear in mind are:

- Highest air pressure is at the point of greatest safety; lowest pressure at the point from which the first steps towards escape must be taken.

- Lift shafts, lobbies, ducts and lavatories on escape routes can cause diversion of air flow. Air input and pressure must be adjusted accordingly (see fig. 15)

- The fans supplying air under pressure must be duplicated with standby units all with automatic control and manual override.

- Air turbulence and contra-flow is created by the opening and closing of doors; external micro-climatic conditions can disrupt exhaust. Air pressure and velocity must be capable of overcoming these problems.

- Doors must swing against the direction of air flow (which will also be in the direction of escape). Closers will not be necessary, except where pressures are very low.

Automatic control of doors and shutters

Smoke control and containment of fire is most normally achieved by arrangements of doors and shutters such that they are held in a closed position when not in use. There are, of course, a great number of circumstances where doors are required to be kept open until the moment a fire occurs, otherwise they may interfere with working operations, sales activities or circulation in general. The range of door and shutter types and the electrical or mechanical equipment operating them is wide; in fact almost any combination that might resolve the designer's difficulties can be fabricated.

Control of a fire door's motive or release mechanism will be by one of the following:

- A fusible link responding to local heat, so as to release a counterweight system or interrupt an electric circuit operating magnetic catches.

- A solenoid switch activated by a local heat or smoke detector or by a general detection and alarm system, so as to set in motion electrically motorised machinery.

All doors that are not in normal use must of course be capable of manual operation. The following are some of the doors available, the manner of their operation and circumstances in which they might be used.

- Automatic sliding doors (usually glass) – Beam or mat-switch controlled, remain closed until approached and therefore remain an effective barrier to smoke; high quality installation used in airports, hotels, shops, etc.

- Hinged doors – Held open by a magnetic catch, released by interruption of electric circuits by fusible link or solenoid switch, closed by standard door head or floor closer; used widely in hospitals, offices, old persons' homes, etc.

- Slide folding steel doors – Motor operated types have solenoid switches activated by the alarm or detection system, manually operated types have emergency counterweight closing device on fusible link. Doors can be manufactured to four hour fire rating and are therefore useful in compartment walls; used mainly in industrial situations where doors are in constant use and the floor track will be kept clean.

- Vertical sliding steel doors – motor operated as above, with solenoid switch or gravity action with counterweights and fusible link. These doors are reliable and can be manufactured to four hour fire rating and used in industrial and commercial situations (e.g. supermarkets, shopping malls, etc.) where they will be held open at all normal times.

- Horizontal sliding steel doors – Very heavy duty and therefore motor operated with solenoid switch control. Advantage lies in their high potential fire rating but disadvantage in their need for a floor track which might not always be kept clean. Used in industrial or commercial situations where there is no head room for vertical sliding doors and where they will be open for most of the time.

- Roller shutter doors and escalator shutters – Normally motor operated with solenoid switch control activated by detection or alarm system – manual operation is normally by chain purchase on the roller spindle. Fire rating is limited by size but roller shutters are useful in all situations where there is a shortage of head room or floor tracks are a disadvantage and where intruder security is not important.

- Overhead doors – Generally lightweight and used for domestic garage purposes, having little fire protection. In industrial circumstances they would normally be motor operated with emergency cut-in solenoid switching.

All electrically operated doors must be on a standby electrical supply circuit if this is provided. The only doors that will function at the time of complete electrical failure are those restrained by magnetic catches and those that operate on counterweight systems (provided the fusible link is broken). It must be restated that completely automatic systems are not acceptable and that manual operation should be possible. Occupants might be cut off from escape by doors that have closed; a crticial factor in the public areas where heavyduty doors are used. Small personnel doors built into heavy doors of overhead or vertical or horizontal sliding type would allow this danger to be avoided.

Wet and dry risers

Wet and dry risers are installed to bring a supply of water close to the seat of a fire so that it may be tackled from inside the building. They are appropriate to a wide variety of buildings or particular areas of buildings where sprinklers or other systems might prove unacceptable or ineffective in protecting life, for example, schools, hospitals, auditoria, small hotels, offices and other commercial or industrial situations. They are suitable where all carbonaceous fires are involved, e.g. wood, paper, textiles and furnishings.

Wet risers and associated hose reels

For their simplicity and efficiency these are preferred where the existing water pressure is sufficient for their proper operation, (in accordance with the F.O.C. rules and the Building Regulations).

A fully charged water supply pipe has a hose reel permanently connected to it. When the inlet valve on the supply pipe is opened, the hose reel is immediately ready for use; the nozzle is removed from the bracket and the hose run out towards the fire. The nozzle can then be opened to give a 9 m jet or a spray. Ideally, the water supply for hose reels should be separate from, and independent of, the general water supply to the building. The water mains pressure should be sufficient to provide a running pressure at the nozzle of 0.84 kg/cm^2 (12 lb per sq. in.), or a flow rate of 23 litres (5 gallons) per minute for a range of 6 m (20 ft) at the highest and furthest point. Whilst this standard will comply with the requirements of the Fire Precautions Act and satisfy the FOC, the Greater London Council requirements are more onerous, and in buildings controlled by section 20 of the London Building Acts, a pressure of 2 kg/cm^2 (30 lb per sq. in.) will be required at the highest reel with a flow rate of 136 litres (30 gallons) per minute.

Hose reel installations should conform to the appropriate recommendations of the Code of Practice BS 5306 Part 1: 1976. This calls for one reel for every 800 m^2 of floor area (or part thereof), located in such a position that the nozzle can be taken within 6 m of each part of the area covered. Wet risers and reels should be positioned so that the hose, when run out, will not impede the means of escape or prevent the proper closing of smoke stop doors. They should not therefore be placed on staircase landings. Siting should be on every floor level, adjacent to the point of exit and very obvious. The reel, which can swing out at an angle to the wall, may be recessed so as not to obstruct the corridor or escape route.

Dry risers

These are installed mainly in multi-storey commercial or residential buildings where large quantities of water are required under very high pressure, but where the existing mains pressure is inadequate to reach all levels and where use is to be strictly limited to that of the fire brigade.

A heavy duty uncharged service pipe is provided to all floors from an obviously accessible coupling and valve, ready for connection with a fireman's hose to a high pressure (usually mobile) supply. Once connected, the pipe is charged by opening the valve, the air being cleared from it via a high pressure valve at its head. Single or smaller double couplings with gate valves may be provided at each floor to which firemen's hoses are connected and run off to the location of the fire. Since the hoses (which may be of almost any length) have to be taken to the level of the fire it is wise to locate the dry riser close to a lift under firemen's control.

The virtually unlimited length of hose which might be used means that, in most situations, only one dry riser is necessary in a building for water to be taken to almost any point. Additional dry risers may be provided for more simple strategic reasons, for example difficulties of fire brigade access at ground level, extremely extended floor plans or the intervention of compartment walls with automatically closing doors isolating wings of a building. It will be necessary for the designer to consult the fire officer on such matters, beyond which the rules of location are similar to those for wet risers.

Hand held extinguishers and other equipment

Beside the full gamut of built-in protection systems, the small hand held fire fighting equipment that should be provided may seem little more than incidental paraphernalia. This is not so. Although it must be insisted that priority goes to the evacuation of a building, widespread fire can be prevented by immediate action on the part of occupants who are on the spot and who can find a suitable extinguisher ready to hand. The types of extinguishers which must be made available for emergency include:

- Water extinguishers (water under pressure)
- Foam
- Carbon dioxide
- Halon gas (BCF and BTN)
- Powder.

All are contained under pressure in cylinders of assorted size, ranging from total discharge water, foam and powder extinguishers for general purpose (usually hung low on walls because of their heavy weight) to the latest Halon mix extinguishers which, like standard aerosols, are of controlled discharge and very compact.

The criteria for the choice of extinguishing agent are of course exactly as previously described for in-built systems. All containers describe the contents and the appropriate use (see section on Powder systems above).

In a slightly different category mention must also be made of fire blankets. Originally made of asbestos but now of heavily woven glass fibre, these are rolled tightly and contained in open-ended cylinders from which they can be quickly pulled in an emergency. These are used for class A and B fires and are particularly useful in dealing with cooking fires.

In many circumstances the provision of hand equipment is left to the building user at the direction of the local fire prevention officer after the building is completed. However, this procedure invariably results in difficulties in finding suitable locations or ones in which the appliances are not a physical or visual intrusion and consequently later damaged or moved. Proper positions are:

- In staircase and entry lobbies leading to primary circulation routes –

so that they can be collected on the way in to attack a fire.

- Adjacent to escape doors – so that they can be used on the way out.

- At focal points in work areas and circulation routes – intermediate resources seen from more than one direction.

- Adjacent to special hazards – e.g. electrical switchgear, machinery of all descriptions, etc.

- Outside stores of dangerous combustibles – chemicals, liquid fuels, organic materials liable to spontaneous combustion, etc.

- Adjacent to special work stations – in laboratories, many types of manufacturing facilities, cooking ranges, etc.

In key locations fire extinguishers are best sited en suite with other fire fighting and protection equipment, in specially designed and well displayed fire fighting stations. The outlets for dry and wet risers, together with their hose reels, are available in neatly designed units which, if required in the building, would form the basis for such fire fighting stations. Other items of equipment that might also be included are:

- Fire alarm and call point

- Emergency telephone

- First-aid supplies

- Stretchers

- Protective clothing

- Breathing apparatus

- Portable escape apparatus such as ladders, chutes, etc.

Maintenance of extinguishers and the other equipment listed above can be an onerous task in large buildings or complexes. Plans should therefore be available which show the location of the fire fighting stations with schedules of the appliances giving detailed records of their servicing. Assistance is provided on this matter by the Fire Officers' Committee publication of 'List of Approved Portable Extinguishers 1983'. Not only does this provide approved appliances and names of manufacturers, but also relates maintenance to BS 5306: Part 3 and provision of identification labels affixed to appliances with the distinguishing letters F.E.T.A. (Fire Extinguishing Trades Association, 48A Eden Street, Kingston-upon-Thames, Surrey).

'. . . someone to shout FIRE . . .'

Chapter 5
DESIGN, INSURANCE AND THE F.O.C. RULES

Annual expense of fire insurance

The design of a building should impose on the architect the duty to provide a structure which not only complies with all statutory regulations but will give the user the benefit of economic maintenance and cost in use. Considerable thought is usually given to this subject and much time is spent on balancing capital expenditure with maintenance cost on items such as heating installation, windows, double glazing, self finish surfaces and the use of many maintenance-free materials. However, little time is devoted to the cost of fire insurance premiums, which recur annually and are ever increasing. Architects may suppose that they have no control over this matter; this is untrue.

Fire insurance premiums on buildings can be as low as 0.1% of the sum at risk. Where the risk of fire is great this premium can increase many-fold and poor design can be significant in this assessment.

Statutory requirements and insurance objectives

Most statutory requirements are primarily directed towards safety and preservation of life; they influence the design of the building by reducing the possibility of fire occurring and, in the event of fire, by restricting its spread and providing adequate means of escape. The insurer is concerned mainly with the protection of the property from destruction or damage by fire and the preservation of capital and material assets. One of the important factors which determines the premium rate of fire insurance for a building is the design of the structure to restrict damage in the event of fire.

There are other factors which greatly influence the premium rate. The use, structural condition and location of the building will all be taken into account by the insurance surveyor when preparing his general appraisal and assessment of the risk that his Company will be underwriting. The history of the insured in respect of past insurance matters will also be a consideration of the underwriter. Whereas matters of pure insurance are outside the scope of this book, the design of the building is the responsibility of the architect and the fire risks to the material asset should have serious consideration in his basic concepts.

The attention of the designer must also be drawn to the inadequacy of the Building Regulations regarding fire precautions and consequently the cost of fire insurance premiums. This can best be achieved at design or construction stage of the building. So far as the Building Regulations are concerned, no difference is made between a warehouse housing polystyrene goods and another storing steel beams; yet the fire risk is entirely different and unless the design and construction take this into account the fire insurance premiums can be penal.

Insurance premiums are related to risk. In respect of buildings, the fire risk relates to the material asset. Premiums for fire insurance of buildings are the charges for the considered business risk undertaken by the insurers relating to the value of the material asssets at stake. Risk improvement therefore is a positive way for the designer to reduce vulnerability to fire and consequently to reduce fire insurance premiums.

Rules of the Fire Offices' Committee

The whole matter of fire and building design is complex. From inception, or acceptance of the brief from the client, the designer of the building will be greatly influenced by the planning laws, Building Regulations and local authority Acts and byelaws. He must also apply himself to understanding the Rules of the Fire Offices' Committee and take account of the cost in use of complying with these Rules in the design of the building.

The Fire Offices' Committee operates from Aldermary House, Queen Street, London EC4P 4JD and it is from there that all their publications can be obtained. The Committee, usually referred to as F.O.C., is a body which was founded by and is comprised of representatives from most of the major fire insurers in the United Kingdom. Originally these were what the insurance market calls the 'Tariff Companies' but their Rules have always been very well respected by almost the entire fire insurance market. They are used as a fundamental fire insurance basis for the whole underwriting world including many Non-Tariff Companies both in the United Kingdom and from abroad. These Rules are not part of the Building Regulations or Acts or any other statutory requirements. They form one of the basic considerations in determining the premium rate which is to be charged on a building for fire insurance. Designers should be fully aware of these Rules at the initial design stage as the fire hazard of a building will be rated by the underwriters according to important factors such as type of construction, nature of occupancy and management. Premium savings are often lost on the drawing board because the F.O.C. Rules are not considered in the design and by the time the insurers review the fire risk on the building the requisite alterations prove impossible or are cost prohibited.

The F.O.C. Rules are constantly under revision and there are many Reissues, Amendments, Addenda, Approved Lists, and Recommendations. It is therefore very important to confirm that the information on which a design is based is the latest. The Rules which will primarily affect the fire rating of the building are in the following publications:

- Rules for the Construction of Buildings: Standards I–V.

- Rules for the Construction of Buildings: Grades 1 and 2.

- Rules for Automatic Sprinkler Installations.

- Rules for Automatic Fire Alarm Installations.

- Rules for the Construction and Installation of Firebreak Doors, Lobbies and Shutters.

As these publications are very precise and exacting it is not intended to give a précis of their contents. A brief indication of the scope of each document will, however, introduce the subject matter.

Rules for Construction of Buildings: Standards I – V

Each Standard clearly sets down the limits and requirements of each structural element in a building to enable it to conform to that particular Standard. Standard I has the most stringent requirements and Standard V has the least stringent requirements; the Standards between are graded accordingly. Should a building not fulfil the requirement of at least Standard V, then the building is regarded as Non-Standard. Non-Standard buildings are liable to attract an increase in basic or normal fire insurance rates, while Standard buildings will attract basic rates, or a discount if they conform to one of the more stringent Standards.

The general effect of these Rules is to divide a building into fire compartments and one of the ways to limit the loss that can occur through fire is to reduce the total area at risk. This works in two ways. First, by limiting the area, the development of heat is restricted making it less dangerous to surrounding risks and also easier for fire fighting services to limit and extinguish the fire. Secondly, there is the actual restriction of the spread of fire into other parts of the same building. 'Fire Compartmentation' has previously been described in chapter 3 and is the division of large hazardous areas and large buildings into separate fire compartments by means of fire resistant or 'firebreak' walls, floors, and ceilings. Proper isolation by fire compartmentation of high fire risk areas from normal or low risk areas results in reduction of premiums.

A designer, by making a small change in construction, can change a building from Non-Standard to Standard or from say Standard III to Standard II. This might be effected by a correct decision on a relatively optional choice of a material or by increasing the thickness of a wall or floor by say 5 mm. Alternatively it might be achieved by providing scuppers at very little cost in external walls to facilitate the removal of water during fire fighting; or by eliminating a fanlight above a door. It is the capital asset that insurers wish to preserve and where this is achieved by conforming to the Rules for a particular Standard the fire damage risk is reduced and the fire insurance premium will be lower.

In 1979 the F.O.C. agreed that these Rules (i.e. Standards I – V) need only apply to buildings erected before 1 March 1978. Designers should note that this applies to the majority of building stock in the United Kingdom and when carrying out any refurbishment or alteration to existing buildings the fire insurance premium is likely to be assessed in accordance with these Rules.

It is important to note that where a Standard of the F.O.C. has been achieved it must not be assumed that the statutory or legal requirements have been fulfilled. Neither must it be assumed that if the statutory or legal requirements are fulfilled the building will necessarily meet a Standard that has been set by the F.O.C.

Rules for the Construction of Buildings: Grades 1 and 2

One of the major considerations taken into account when determining insurance premiums is the use to which the building is to be put. In 1972 the F.O.C. introduced Rules for the Construction of Buildings: Classes 1, 2 and 3. These Rules applied to premises used for the manufacture, processing and storage of plastic materials. The F.O.C. subsequently regarded the Rules and called them Grades 1 and 2. At the time these Rules were issued, the F.O.C. indicated that it was intended that they would apply progressively to buildings in other occupancies as well as to the plastics industries. In July 1979 the F.O.C. announced that these Rules could be applied to any building erected after 1 March 1978, regardless of occupancy, and designers were advised to consult insurers at planning stage.

The Grades 1 and 2 are prepared with precise and exacting requirements in a manner similar to Standards I – V. As a general guide, however, Grade 1 construction defines a building of non-combustible and fire resisting construction based on two hour fire resistance, and Grade 2 construction defines non-combustible construction based on half hour fire resistance.

Some new terminology was used and 'firebreak' walls and 'firebreak' floors were introduced. If a building does not comply with either Grade 1 or 2 it will be a Non-Graded building and may be subject to premium surcharges. A Grade 2 building should not incur such a penalty and a Grade 1 building can qualify for insurance premium rebate or discount.

Rules for automatic sprinkler installations

In some buildings it is a statutory requirement to install sprinklers. Occupiers of other buildings in which sprinklers are not required by law may experience great difficulty and expense in arranging fire insurance cover. In such cases, the capital cost of installing sprinklers may well be offset by large discounts on fire insurance premiums equivalent to 100% return on capital within a short period of three to five years.

An automatic sprinkler installation provides:

- Detection

- Warning

- Containment

- Extinguishment (in certain cases).

The F.O.C. Rules which govern the correct installation of sprinklers have been varied over the years and are identified by the number of the Edition of the Rules. The current Rules are the 29th Edition, subsequently amended and revised. These Rules specify the minimum Standards required for installation of sprinklers in buildings under the following general classifications:

- Extra light hazard – Usually non-industrial occupancies such as houses, flats, hotels, hospitals, museums and schools.

- Ordinary hazard – Usually industrial and commercial occupancies involving use of materials unlikely to develop instantly burning fires in the initial stages. This hazard is divided into four groups, each group demanding a higher standard of sprinkler installation depending upon the use to which the building is put.

- Extra high hazard – Industrial occupancies with exceptionally high fire loads, either by virtue of special pile storage or because of the presence of materials with potentially rapid burning rates.

The designers should study the detailed requirements in these Rules at an early stage and take advice from insurers as to their effectiveness and their cost in use. Many considerations will have to be given as to how and when sprinklers should be specified and the designer should give early thought to:

- Legal requirements.

- F.O.C. requirements.

- Insurers requirements.

● Availability of water supply to give correct pressure and flow rate.

● Effect of the saving on the fire insurance premium related to:
  Building structure
  Plant
  Raw material stock
  Manufactured stock
  Consequential loss policy.

● Possibility of writing off the capital cost of sprinkler installation as plant against profits in the first year of installation.

● Effect of an extension or alteration to an existing building. Unless an extension can be completely isolated from existing parts by 'fire-break' floors and walls, then the new and old will be rated as one risk and that risk will be based on the worst condition that prevails.

Rules for automatic fire alarm installation

These Rules are prepared on lines similar to those for automatic sprinkler installations, however because of their nature they are less exhaustive but just as exacting. The current issue is the 11th Edition, with various amendments. The Rules apply particularly to heat and smoke detectors and cover such matters as definitions, classification of installations, areas to be protected, siting and spacing, apparatus generally included, location, power supplies, wiring and connection and testing and maintenance. These will influence insurance premium discounts.

The need for and advantages of automatic fire alarm installations will be obvious to all designers and their early consideration to this matter should be in line with our recommendations for sprinklers. There is no doubt that the best warning of fire is the early warning. The efficiency of the fire brigade is not judged by the gallons of water it can pump in a minute but by the minutes it takes the fire-fighting appliance to reach the scene of the fire.

The designer has available a variety of warning systems that can be installed in either a new or existing building in order to detect fire. Detailed examination of these installations has been made in chapter 4. The warning systems can alert the occupants of the building and also the fire brigade. Fire detectors are designed to detect one or more of three characteristics of a fire, namely smoke, rapid rise of temperature or flame. No single type of detector is the most suitable for all applications and the final choice has to depend on individual circumstances. In some premises it may be useful to combine different types of detectors or to install both a detection system and an extinguishing system, such as sprinklers. In other circumstances, statutory obligations may require the designer to install alarm or warning installations before a Fire Certificate can be issued.

Warning can be given either manually or automatically. The first reaction to fire is for someone to shout 'Fire'. In many smaller types of building this is acceptable as the best method for warning. Fire bells that can be manually activated by the occupants range from the hand bell to the push button types which sound alarms throughout a building and raise direct assistance from the fire brigade. Flashing lights which are also manually activated are more appropriate warnings for deaf persons; where the designer is concerned with disabled people he must give very special attention to the provision of special visual or audible warnings.

The F.O.C. provides a list of approved automatic sprinkler equipment and a list of approved automatic fire alarm systems, together in each case with names and addresses of suppliers and installers.

Rules for the construction and installation
of firebreak doors, lobbies and shutters

The scope of these Rules is described in the heading. They comprise four sections as follows:

Section 1   Doors and Shutters made to a specification issued by the Fire Offices' Committee.
Specification 1: Iron or steel doors.
                2: Metal covered doors.
                3: Steel rolling shutters.
These three specifications give requirements in respect of:

- Single firebreak doors.

- Double firebreak doors.

- Wall opening; detailing maximum size, type of sill, jambs and head, floor covering, lintel and clearances.

- Certification of door.

- Construction of door; detailing measurements and materials for plates, rails, styles and fixing of bolts, straps and other fittings.

- Hinges and pivots.

- Bolts and latches.

- Frame.

- Sliding door details.

- Rolling shutter details.

- Lists of approved manufacturers.

Section 2 Proprietary Doors made to a manufacturer's own specification.

This section provides a list of approved manufacturers and types of doors and provides an F.O.C. reference number, type of construction and fixing (whether hinged, sliding or folding) and maximum size of wall opening.

Section 3 Firebreak Lobbies.

This section gives a specification for the construction of firebreak lobbies.

Section 4 Firebreak Shutters for the protection of openings in external walls.

This section refers to shutter protection to windows or other similar openings in external walls and the specification is related to sections 1 and 2.

Architects beware!

It is not our intention to detail the basis on how a building is rated for fire insurance purposes, but it is clearly our intention to illustrate that there is a very definite and exacting procedure that is followed by the insurer for this purpose. The architect should make himself aware of this procedure by obtaining copies of the proper documents and by consulting insurers, so that consideration is given to this matter at early planning and design stage.

The designer must carefully weigh up the advantages and costs of escape against security; compartmentation against open plan; ventilation ducting and suspended false ceilings against providing a vehicle for spread of fire and smoke. He must think 'FIRE' at design stage and ensure that the client not only gets the best return on his capital but also a building with a low fire risk and low fire insurance premium.

'. . . *maintenance of escape routes*. . .'

Chapter 6
OCCUPATION AND THE FIRE CERTIFICATE

Under the Fire Precautions Act 1971, it is unlawful to occupy a building – new or old – for certain designated uses without having, or having applied for, a fire certificate issued by the fire authority.

Fire certificates relate to occupancy and although they apply to the use of new as well as old buildings, in the case of new buildings or extensive alterations compliance with the Building Regulations will have covered the structural requirements of the fire authority. Conflict between the requirements of the Building Regulations and the fire authority has to be avoided and this is covered by the 1971 Act.

What is a fire certificate?

To make sure that life is protected and people can escape safely and quickly when fire breaks out in a building, the Fire Precautions Act 1971 empowers the fire authorities to examine all premises within the scope of the Act and to issue a certificate once they are satisfied that the means of escape and related fire precautions are properly provided and are likely to remain effective in an emergency. The fire certificate is the weapon of enforcement of the Act.

The Act requires that a fire certificate has to be obtained for all premises before they are put to a use or uses designated in an Order under the Act made by the Secretary of State for the Environment. In principle, these are uses of premises where people are present in large numbers, or in other circumstances where they would be specially at risk in the event of fire. It is unlawful to occupy and use a building for any of these designated uses without a fire certificate.

The Secretary of State may also make regulations applying the fire certificate provisions to other uses in places which are not necessarily permanent buildings, such as moored and beached craft, booths, circus big tops and similar moveable structures if used for a purpose where the public may be at risk.

In addition, the Act empowers the fire authority itself to serve a notice requiring particular premises used as a dwelling to be covered by a fire certificate.This procedure is adopted where the fire authority considers there is a greater risk to the occupants than in an ordinary dwelling house, such as housing for the elderly.

This greater risk is defined as living accommodation which is:

- Below the ground floor.

- Two or more floors above the ground floor.

- Six or more metres above the surface of the ground on any side of the building.

- In a building where explosive or highly flammable materials are stored.

Designated uses

It is reasonable to assume that in due course all classes of use set out in the Act will be covered by a fire certificate, but, at the time of publication, Orders under the Act have only been made for three uses, namely:

- Hotels and boarding houses – If there is sleeping accommodation for more than six guests or staff; or if there is some sleeping accommodation below ground floor or above the first floor.

- Factories – in which more than twenty persons are employed to work at any one time; or

  in which more than ten persons are employed at any one time elsewhere than on the ground floor; or

  which are in the same building as other factory, office, shop or railway premises and the sum total of employees in all premises exceeds the above; or

  if explosive or highly flammable materials are stored or used in or under the premises and constitute a risk to persons in the premises in case of fire.

- Offices, shops and railway premises – Numbers of employed and locations are the same as for factories.

Each of the Orders made is accompanied by the issue of a Guide in two parts published by H.M.S.O.

Part I is a guide to the Fire Precautions Act 1971 as it applies to the particular designated use and covers:

- Definitions, application and issue of a fire certificate, including necessary improvements before a certificate is issued, obligations and changes of conditions and alterations to premises.

- Rights of appeal.
- Effect of the Act on other legislation.
- Regulations under the Act.
- New buildings.
- Consultation.

Part II sets out the fire precautions required and covers:

- Fire resistance and surfaces of walls and ceilings.
- Assessment of risk.
- Means of giving warning.
- Fire fighting equipment.
- Fire instruction and drills.

Exemptions

Fire certificates are not required for the following premises due to special exemption under the 1971 Act:

- Premises for public religious worship. It should be noted, however, that regulations may be made specifying fire precautions for churches and other places of worship and that church halls used for a variety of purposes are unlikely to be exempt.
- Houses occupied as single dwellings.
- Houses in multiple occupation.
- Prisons and similar penal institutions and certain special hospitals.
- Premises occupied by the armed forces of the Crown, as distinct from premises owned or occupied by the Crown.
- Certain special premises such as nuclear installations, explosives factories, mines and large petroleum installations which are under the control of the Health and Safety Executive.
- Certain office, shop and railway premises to which the 1963 Act, covering these premises, does not apply – such as places where only self-employed people work or businesses where only close relatives of the employer work.

Application for a fire certificate

The application should be made by the occupier to the fire authority. Anyone can make an application but the responsibility remains with the occupier, unless the building is in multiple occupation, in which case the owner becomes responsible. In practice it is advisable for the architect to initiate the application during construction or prompt the occupier or owner to do so. An application exempts the person who has applied from prosecution until such time as he receives either an approval or a rejection. A copy of the Application Form FP1 (rev) is shown in appendix 5.

Issue of a fire certificate

On receipt of the application the fire authority may request further information – in particular, simple outline plans showing the essential features of the building. The fire authority will then inspect the premises and if satisfied must issue a fire certificate. If the authority is not satisfied, a notice will be served on the applicant informing him of what needs to be done before a certificate can be issued and will specify a time limit for the works to be carried out. If the work is not done in that time, subject to further time allowances or appeals, the application will be deemed to have been refused and it will be unlawful to use the premises for the use required.

An example of a fire certificate is included in appendix 6. In principle it will specify:

- The particular use or uses of the premises which it covers.

- The means of escape in case of fire. A plan will often be used for this purpose.

- The means for securing that escape routes can be safely and effectively used at all material times. This would cover such matters (where appropriate) as emergency lighting, direction signs and smoke stopping.

- The means for fighting fire for use by persons in the building.

- The means for giving warning in the event of fire.

The fire certificate may also be used to impose requirements relating to:

- The maintenance of the means of escape and keeping them free from obstruction.

- The maintenance of other fire precautions specified in the certificate.

- The training of members of staff and the keeping of appropriate records.

- The limitation of the number of persons who may be in the premises at any one time.

- Any other relevant fire precautions.

New buildings or extensive alterations

Since fire certificates are required for all existing buildings with designated uses and since new buildings or extensive alterations are designed to comply with Building Regulations, the Act is at pains to ensure that there is no conflict between the fire authority and the local authority. For example, the Act provides that the fire authority (for the purposes of a fire certificate) may not require alterations to be made to a building which has already been subject to the provisions of the Building Regulations relating to means of escape.

Similarly, if a fire certificate is required for a building when it is brought into use, the local authority is required to consult the fire authority before dispensing with or relaxing any requirements in the Building Regulations relating to structural fire precautions or means of escape and before passing plans under the Regulations. The designer of a building may find himself having to comply with local Acts, each with their own requirements for fire precautions. For example, some authorities have their own special Acts relating to sprinklers. Acts such as these may require premises to be licensed and fire precautions may be imposed as conditions of the licence. These Acts continue in force but where requirements relating to means of escape and fire precautions are imposed under them, such requirements will have no effect in the case of premises for which a fire certificate under the 1971 Act is in force. The 1971 Act will thus gradually supersede provisions under other legislation relating to fire precautions as it is applied successively to various classes of premises. An occupier or owner is thus excused from doing anything under a local Act which would involve a contravention of the 1971 Act or regulations made under it.

Phased completion contracts

Assuming that a building contract involving phased completion also involves phased occupation it follows that a fire certificate will have to be applied for and issued for each completed phase before it can be put to use. Thus when planning the building it would be sensible to ensure that each phase complies with the fire authority's requirements and that a certificate is likely to be granted for that part of the building.

Legally a building can be used once an application has been made for a fire certificate and before a decision has been given.

However, the insurance company will also be involved and is unlikely to be sympathetic to a claim arising during a time when a partially completed building is occupied without a valid fire certificate. The insurers may argue that the application could eventually result in a refusal.

Clearly, in phased occupations, close liaison with the fire authority, the insurance company and the intending occupier or owner is essential. Discussions should begin at the planning stage and agreements should be reached well before the date of handover.

Speculative building

It is not normal but it is legally possible to obtain a fire certificate for a speculative building before a tenant user is known and before the building is occupied.

Validity period

A fire certificate remains valid indefinitely, so long as there are no changes in the circumstances. The onus to notify changes in personnel employed or re-arrangement or alterations to the premises, including any change of use, remains with the occupier (or owner when in multiple occupation). The fire authority has the power to inspect under the Act but there is no provision for regular inspections to take place. The frequency of inspections is determined by the fire authority and varies throughout the country.

Maintenance of means of escape

Continuing maintenance of the means of escape is the responsibility of the building owner and penalties can be imposed by the Courts in cases of default. Maintenance requirements are set out in the certificate. The following is a quotation from the GLC Code of Practice:

'The lives of persons can be seriously prejudiced if, in fire conditions, for instance, the protection afforded to escape routes is damaged, or fire resisting self-closing doors are fastened back in the open position or are wedged open and do not close tightly against their frames. Applicants acting on behalf of a client should ensure that:
- the owner and/or occupier is made fully aware of his responsibility in this respect

- all occupants of the building are fully acquainted with the routes of escape and final exits

106

OCCUPATION AND THE FIRE CERTIFICATE

- the owner and/or occupier establishes trained key personnel on each storey of the building with a view to the speedy marshalling of the occupants and the orderly evacuation of the various storeys should the need arise.'

'a dry riser'

Chapter 7
MANAGEMENT

There have been a number of major disasters which have drawn attention not only to the need for buildings to be adequately designed and protected against fire but which have also highlighted the responsibilities of the designer and the manager of the building in relation to fire safety. In particular they have focussed on the role of the building manager in maintaining a safe building.

Three major disasters, at:

● The Summerland leisure complex, Isle of Man, in 1973

● The Fairfield Home, Nottinghamshire, in 1974,

● Woolworth Store, Manchester, in 1979

were each the subject of enquiries and the reports of the subsequent findings should be compulsory reading for all designers and building managers. What all three fires appear to have in common is that in each case the tragedy arose from the co-incidence of a number of factors rather than a single cause; from a cumulation of individual errors and failings rather than from the shortcomings of any one person.

The major lessons to be learned from these fires are:

● The speed with which fires can develop: following the Woolworth fire, tests carried out at Cardington on a stack of packaging materials, soft furnishings and bedding showed that the fire would have reached maximum severity two minutes after ignition.

● The rapidity with which smoke can spread, making evacuation difficult and hazardous.

● The majority of deaths in fires are caused by smoke: of the eighteen deaths at the Fairfield Home seventeen were caused by carbon monoxide poisoning from smoke.

● The need to maintain the integrity of escape routes: the doors at the foot of the north-east service staircase at Summerland were fitted with panic bolts. These doors were made ineffective by being padlocked and by there being a car parked immediately outside.

Thus the priorities for management must be:

- Regular maintenance of the warning system
- Regular maintenance of the direction and escape signs
- Regular maintenance of fire appliances
- Maintenance of unimpeded escape routes
- Training of staff in evacuation procedures, and
- Training of staff in first aid fire fighting.

The role of management

The designer must have an understanding of the needs and problems of management. The manager in turn must have a detailed knowledge and understanding of the building, its services and its safeguards, to assist not only in the day to day running of the building for its primary use but also to help in any emergency which may arise.

The broad responsibilities of management are:

- The day to day operation and maintenance of the building
- The security of the building
- The maintaining of adequate safeguards in the event of an emergency.

Present day buildings are complex. The variety of building types with which the designer and the manager are likely to be involved is greater than ever before. Both should be fully aware of the causes and behaviour of fire and of how people are likely to behave in a fire.

At the early planning stage the designer should be able to draw on his knowledge and expertise to try to visualise how fire is likely to affect the building and its occupants and plan to confine the fire within containable areas and to provide for easily identifiable routes of escape.

In many cases the client may well be the manager of the building or the manager may be involved from the outset. However it is quite likely that the manager will be appointed after the design process is complete and will not always be aware of the reasons for the various design decisions.

On completion, the building will comply with the current regulations and the requirements of the fire authority. In many instances a fire certificate will have been obtained. On occupation, management will become responsible for day to day maintenance and assume responsibility for complying with the requirements of the fire certificate.

Probably the most important management task will be to maintain the integrity of escape routes and to ensure adequate smoke free means of escape from all parts of the building direct to the open air. In a properly designed and maintained building occupants, including any who are disabled, should have no trouble in leaving quickly and safely. If not properly maintained these 'safe' routes may themselves become death traps.

The main hazards in escape routes are normally the doors as these are so often misused. The commonest door is the self closing type and, used as provided, no problems should arise. But so often doors of this type can be found propped open to allow free access during the normal use of the building. Temporary wedges can often become permanent and in some buildings it is not unusual to find such doors fitted with cabin hooks to keep them open.

Corridors leading to protected staircases are often used for temporary storage purposes – usually furniture but sometimes files and other bulky items. The result is at best to reduce the width of the corridor and at worst to block the means of escape altogether.

Potential storage areas may also be found at the foot of escape stairs; not only can carelessly stacked items be likely to block the exit doors but they may constitute a high fire risk, turning a clear route to the outside into a trap.

Exit doors fitted with simple panic bolts or latches are often secured with chains and padlocks to prevent illegal entry. In such circumstances they will also prevent egress in an emergency.

Outside the building, access routes for fire appliances must be maintained unobstructed and not allowed to become car parking areas or stacking places for surplus materials, etc.

The tragedy is that the effect of such misuse is rarely obvious until an emergency arises.

Security

One of the biggest problems facing the manager is that of maintaining adequate security. On the one hand is the need to ensure that the building can be evacuated safely; on the other hand is the need to prevent entry to the building by unauthorised persons. Conversely, special provisions must be made in buildings such as prisons and mental institutions, where containment of the inmates remains a factor even in the event of fire.

Crime is a particular problem. Obvious high risk buildings like banks and similar institutions are now so well protected that robbery has turned its attention to more vulnerable areas, in particular picking-up points for cash at supermarkets, dairies, shopping precincts, etc. The positioning of secure rooms for the purpose of cashing up, making up wages and the storage of high value goods needs to be carefully considered with both robbery and means of escape for the staff in mind. In purpose-designed buildings this should be relatively simple. Where problems are bound to arise is in the fitting out of building shells or in subsequent alterations.

Vandalism has to be considered, especially arson. It is perhaps significant that, currently, nearly one third of all fires result from arson, which is difficult to prevent. The fire at Summerland, with the loss of fifty lives, was started by boys setting fire to part of a disused fibreglass kiosk which had been left lying on its side close to one of the external walls.

The resolution of the seemingly incompatible problems of emergency escape and security will best be achieved if entrances, exits and security areas are considered at the earliest stage. Entrances should be kept to the minimum necessary for the effective use of the building and emergency exits should be planned to be used solely for that purpose, with suitable quick release security locks. Dual use entrances/exits should be kept to a minimum, when the problem of achieving control of the entrances in use and the security of the building when closed will be much reduced.

Management must not only understand the building and make sure that staff are trained in fire routines and evacuation procedures but must also ensure that regular checks and maintenance are carried out on mechanical aids and fire fighting appliances, whether fixed or portable, and the results entered in a manual as a continuing record.

The building manual

The task of management will be greatly helped if all the information relating to the building is available for reference. The initial responsibility for preparing this information ought to belong to the designer. Probably the best way to bring together the diverse information necessary for the effective management of a building is to prepare a building manual or log which will eventually chart the history of the building from its initial design to its ultimate demolition. This manual, as issued on completion of the building, should contain the client's brief and full information relating to the building and eventually a complete history of all adaptations.

An important part of the manual, usually required by the terms of a fire certificate, is that covering fire safety and means of escape. These aspects can probably best be illustrated graphically using plans of the building.

Site/location plans showing:

- Vehicle access for cars, lorries, service and emergency vehicles.
- Emergency escape routes.
- Hard standings for emergency vehicles.
- High risk stores for petrol and other flammable liquids, gas and liquid gas, toxic materials etc.

Plans of each floor showing:

- Services: positions and runs of water, gas, electricity, drainage, with all controls clearly indicated.
- High fire risk areas: boiler rooms, kitchens, laboratories, storage areas, identifying flammable and toxic materials, switch rooms.
- Emergency generators.
- Emergency secondary lighting: identifying power source and type.
- Escape routes: identifying all fire, smoke and escape doors with direction of opening and types of ironmongery, all positions and wording of signs.
- Lifts: indentifying those for fire brigade use.
- Fire alarm systems: whether manual, automatic, local or linked to the public emergency services.
- Fixed fire fighting installations: smoke/heat detectors, dry and/or wet risers, sprinklers, hose reels.
- Portable fire fighting appliances: positions and types, tools, extinguishers.

From hand-over, the responsibility for the upkeep of the manual will rest with the manager who should ensure regular monitoring and updating of:

- Plans: ensuring that they incorporate any changes in structure, layout, usage, fixed and portable equipment, services, etc.
- Fire drills: noting frequency with any observations.
- Equipment tests: ensuring that all fixed and portable equipment and installations are serviceable,
- Training: ensuring that staff are fully conversant with fire fighting and evacuation procedures.

● All necessary licences and certificates.

The building manual is an important document and, together with the fire certificate, should be kept secure but accessible. The designer should pay particular attention to the preparation of the manual and should impress on the manager its value as a management tool.

Alterations

Buildings change hands from time to time and very often their use may change. Change usually brings about alterations, from the ad hoc alteration done by the works manager to cater for a local change in use, to a total refit. If any alterations are attempted without a knowledge of the original brief and design philosophy, potential hazards may be brought into the building which may not become apparent until disaster strikes. Indiscriminate alterations without a full understanding of the possible implications of such modifications can produce potentially dangerous situations; particularly if these are attempted without the assistance of qualified professionals.

No alterations, however innocuous they may seem, should be made without first consulting an experienced fire prevention or safety officer and, equally importantly, the insurance company. Alterations to the building may involve the shuttingdown of a sprinkler system which will require a dispensation to be obtained from the insurers for the period involved. In any case the fire certificate must be amended or a new certificate issued as may be required.

Maintenance

Repairs and replacement of materials must, particularly where they affect means of escape, be of the same grade. Fire resisting glazing must be made good to the original standard, intumescent ventilating grilles and smoke seals properly replaced and flame retardent linings and curtains maintained.

Fire fighting

Designers should be aware that while staff will not be expected to cope with fires beyond the early stages they will be responsible for the safe evacuation of a building. The correct positioning of static hose reels and portable appliances therefore, together with easily identifiable escape routes, will be vital since the regular fire brigades must not be relied on to assist in the evacuation. They may be involved with another fire or disaster or their access may be impeded elsewhere.

Locating the fire

When the fire brigade arrives the source of the fire can usually be readily identified in the case of a small or simple building. In a large building, or in a building complex, it will be helpful to have a zone locator, together with a set of key floor plans at every entrance in order that the fire chief can quickly identify the seat of the fire. Speed is the essence of fire fighting and the ability to be able to pinpoint the source of the fire and deploy resources quickly and effectively may make all the difference between local damage and total loss. It will be helpful to deposit a set of plans with the local fire brigade. These should show a key identifying the various areas. The plans must be kept up to date or their value will be lost. This is particularly important in the massive complexes increasingly common today, where access may be complicated by separation of vehicular and pedestrian traffic, where large volumes are involved, where large numbers of people are working or at recreation and where the use to which space is put is frequently changed. The fire brigade will assist in training permanent staff and (of interest to insurers) maintain an actual presence during special occasions.

Chapter 8

SUMS INSURED

Accepting the risk

Once a new building is completed and handed over, or a contract has been signed for the purchase of an existing building, the new owner takes on the responsibility for fire insurance and the type of policy adopted depends on his requirements and attitudes. A landlord or owner-occupier may have some flexibility in choosing the most appropriate form of insurance; a lessee will need to comply with the terms of the lease.

Covering the risk

All insurance is based on the broad principle of indemnity. The insurer underwrites the risk to indemnify the insured up to the value for losses arising from the eventualities set out in the policy. The intention of the policy is to provide indemnification against losses so as to put the insured back into a position comparable to that before the loss occurred.

In what is generally referred to as a fire insurance policy, the prime risk is damage by fire. However, the policy usually covers other risks such as damage by storm, burst pipes, impact and malicious action. Whatever the risks to be covered, one major item to be determined in the policy will be the sum insured. Not only does the sum define the financial limit of the liability of the insurance but it is likely to be used in other negotations. To ascertain the correct and appropriate sum insured the next major item for consideration is to agree the basis of insurance.

Basis of Insurance

Having understood that all insurance provides indemnification for the insured, there is an alternative basis whereby reinstatement clauses can be included in the policy. Consequently, there are alternative methods of assessing the level of compensation for damage suffered, namely:

- Indemnity, and
- Reinstatement.

117

Broadly, the value of a building assessed on an indemnity basis will reflect its age, state of repair and suitability for use, whereas the value assessed on a reinstatement basis will reflect the cost of replacement of that building irrespective of such age, state or suitability. The former will tend to produce a lower figure and the latter a higher one.

Each alternative merits fuller examination and this will follow, but first there is one further principle of insurance practice that must be understood – that is 'average'.

Average

Average is an insurance term and relates the total value of the material asset at risk and the financial loss to the adequacy of the sum insured. Irrespective of the basis of the insurance, the sum should be adequate to meet a total related to the type of cover provided in the policy. Where the sum insured is inadequate, the compensation will be reduced in proportion to the shortfall of the material asset at risk to the sum insured. The formula for compensation will be either:

$$\text{Cost of repair} \times \frac{\text{Sum insured}}{\text{Asset at risk (reinstatement)}}$$

or

$$\text{Value of damage} \times \frac{\text{Sum insured}}{\text{Asset at risk (indemnity)}}$$

An illustration may make the position more clear by taking the cost of repair of a fire loss at £30 000, the sum insured on a reinstatement basis, at £100 000 , and the cost of totally reinstating the building at £250 000. The financial risk of the insurers is limited to the sum insured and therefore the compensation would be:

$$£30\,000 \times \frac{100\,000}{250\,000} = £12\,000$$

The £18 000 balance of the repairs would have to be met by the insured. If the same figures were related to values and indemnity instead of costs and reinstatement the same situation of average exists. It therefore becomes of great importance to ensure that the sum insured is adequate to meet the total reinstatement cost ('reinstatement') or total value ('indemnity').

Overproviding in the sum insured brings no advantage, except to the insurance firm by way of increased premium. Under no circumstances can average be operated in a manner to increase the financial compensation. Average is a protection to the insurance firm to limit their share of the risk by reference to the sum insured.

Indemnity

Indemnity is related to the value of the material asset. In insurance terms it is often assessed by depreciating the current purchase price of labour and materials needed to reconstruct the building by allowing a deduction for wear and tear. Indemnity might therefore be described as the depreciated value of a material asset. If indemnity were applied to a bituminous felt covered roof, the view could be offered that the roof structure has no depreciation whereas the felt covering should be depreciated in line with its remaining effective life. In a similar way this could apply to decorations, electrical installations, heating systems, floor tiling and other items having a limited life. Brickwork, concrete, steel and timbers should not normally qualify for depreciation unless there is evidence of excessive wear and tear or structural failure.

The application of an indemnity valuation to buildings has to be applied to the whole of the building and not to a part. There is, however, a problem of application and this can be simply illustrated. Assume a warehouse has a market value of say £200 000, of which the land is worth £40 000. An indemnity policy to cover the fire risk on a building would therefore set the sum insured at £160 000. If the warehouse were totally destroyed by fire then £160 000 would be the correct sum for payment for indemnity compensation. The land could be sold for £40 000 and together with the proceeds from the insurance policy the insured would have a combined total of £200 000. The owner would then be in a position to purchase a similar type of warehouse. Hence following a total loss, the indemnity arrangement would have placed the insured in a position similar to that before the loss; a 'like for like' position.

However, it is unusual for damage by fire to result in total destruction. More often a relatively small percentage of the building is damaged and repair has to be undertaken. The indemnity cover now becomes a notional situation as it is impossible to carry out repair with materials and labour that are depreciated by wear and tear. The notional repair costs can only be redressed by a financial contribution, being a proportion of the cost of repair using new materials and labour. Building costs now enter the equation. To rebuild the same warehouse would cost say £300 000 and being insured on an indemnity basis the compensation would be in the ratio of £160 000 to £300 000 of the repair costs. The insured would find that almost half the repair costs were uninsured.

The matter could become very severe if the warehouse were completely destroyed and if, because of trading facilities, it became essential to rebuild on the same site thus providing a new building in lieu of one in a poor condition.

Betterment

The difference between the indemnity value of the old materials and the cost of new materials in the above example is called betterment by insurers. Betterment financially quantifies the improvement of using new materials in reinstatement or reconstruction over the depreciated value of the old materials in a building before the fire took place. Betterment can also apply to improvements from redesign.

The example given shows that in determining insurance indemnity compensation, it invariably becomes necessary to determine the building costs of reinstatement. An exception would be when the building is not repaired, leaving the owner to sell the property in its damaged state and to redress his position on market values only. In this situation the financial compensation is unlikely to balance the financial loss and, consequently, relating the sum insured to indemnity value can have severe financial penalties in resolving an insurance claim for fire damage.

Reinstatement

In determining the sum insured for a reinstatement policy the insured is providing to recover the cost of repair, whether the damage be partial or complete. The advice on this sum should come from a quantity surveyor as he is clearly the best qualified person to determine likely rebuilding costs. Where the building is new and the quantity surveyor has been concerned with the construction, the details from his final account will provide current and reliable cost information. In cases of existing buildings where no cost information is available, the quantity surveyor will rely on his professional expertise in arriving at an approximate estimate of the rebuilding costs. The reinstatement cost will incorporate many factors, which together will enable the insured to be reimbursed for his total expenditure in repairing or replacing the building after damage by fire. Whatever has to be paid for the repair of the building will be reclaimable from the insurers provided the sum insured gives adequate cover. Unlike the indemnity basis, new materials will replace old without application of betterment.

The first factor for consideration in calculating the reinstatement sum insured is to determine the cost at a prescribed date of rebuilding the structure with materials and in a style similar to that which now exists. The quantity surveyor will use in his calculations known cost information, building indices, approximate quantities and pricing, floor area or cubic content, estimating techniques related to other known building costs, or any combination of these methods to arrive at the reinstatement value. The figure at that stage is notional as it will relate to a given date, probably the first day of a new annual insurance contract. The notional cost would assume that the building could be replaced on that day without influence by other factors which, in practice, will always impinge on a notional situation.

Apart from the notional costs of replacing the building the valuer must determine whether the sum insured to be stated in the policy automatically provides for such matters as are listed below or whether it has to be adjusted to take account of them.

- Inflation of building costs during the period of obtaining approvals for planning, building regulations and obtaining competitive tenders.

- Inflation of building costs during the period of site clearance and reinstatement of the building. This would be an average inflation estimate for the rebuilding period.

- Inflation of building costs during the period of the insurance policy, which is usually one year. Where small sums are involved this addition is recommended, but when applied to large sums the inflation element should be provided by an escalation clause. This allows the full anticipated inflation throughout the length of the insurance period and relates the increased premium to half the inflation allowance.

- Costs of demolitions and general site clearance arising from fire damage.

- Costs of protecting and supporting adjoining property arising from fire damage.

- Cost of temporary works and protection to insured property.

- Costs of complying in the reinstatement with current Building Regulations. Many existing buildings have unprotected steel frames, unlined roofs and uninsulated cavity walls which, in a major repair or reconstruction situation, would need alternative and more costly treatment. These costs and those that can be defined relating to planning requirements are usually dealt with in fire insurance reinstatement policies by means of a memorandum or extension to the policy called the 'Local Authorities Clause'.

- Provision of planning requirements which will be considered in more detail in chapter 9.

- Fees for Building Regulations applications, in respect of submission of plans and inspections.

- Cost of upgrading existing sprinkler installations. Any reinstatement of sprinkler installations will have to comply with current Regulations and F.O.C. Rules. Where the existing installation is below the prevailing standards, insurers take the view that the additional cost of upgrading in these circumstances is not covered by the provision of the 'Local Authorities Clause' in a policy. It therefore becomes necessary to arrange a separate insurance item where this particular risk occurs, or to have a special memorandum written into the policy.

- Costs related to the tendering climate of the construction industry projected through the period of insurance. The sum insured is always relating building costs to a future period and to a situation where speed of reinstatement is often vital. Costs established during a period of general recession in the construction industry and where urgency of building does not apply can become misleading in determining the sum to be insured if market conditions change. The position should be assessed and an allowance made if applicable.

- Costs of professional fees and disbursements in preparing and arranging tender and contract documents for the reinstatement work and the cost of supervising the work and preparing interim certificates and the final account.

- Leasehold premises usually have provision in the lease for insurance of loss of rent arising from fire damage. This additional sum should be shown separately in the report. This is one of the few items of consequential loss that is usually covered in a traditional type of fire policy. It arises because of the wording of the lease and the loss of rent relates directly to the building and not to the profit arising from its use.

Value Added Tax

Value Added Tax (VAT) does not apply to new construction and this includes a 100% reinstatement after fire; VAT usually does apply to partial reinstatement. However, most valuations for fire insurance purposes need not allow for the addition of VAT because:

- Where the insured is registered with HM Customs and Excise he should be able to reclaim from them any VAT properly charged by the builder.

- Where the insured is not registered with HM Customs and Excise, and VAT arises from a partial loss, this would be claimable from the insurers. If the cost of partial reinstatement plus the VAT charge exceeded the total cost of reinstating the whole, it would be appropriate to demolish the whole and start again where no VAT would then arise.

There is an area of valuation where the addition of VAT may be applicable. It is usual for a tenant to fit out an open factory or office space with partitions, heating, lighting, computer, rooms, works offices and the like. The responsibility for insurance of these items would be with the tenant. The landlord would insure the structure and two insurance policies would operate. In practice, the tenant may be legally responsible for both insurance premiums, but each policy would function independent of the other. In the event of fire, damage might occur in such a manner that the whole of the fitting out work was lost but with partial damage only to the structure. In these circumstances the Custom and Excise could regard the damage as partial to the whole building and a situation would then arise where the whole of the fitting out work was subject to addition of VAT. Separate valuations of tenants fitting out work for fire insurances purposes where the insured is not registered should therefore provide for the addition of VAT.

'Services below ground'

Chapter 9
PREPARING A VALUATION REPORT

Basis of policy

Before a valuation is prepared on a material asset such as a building, it is essential that the basis of the insurance policy is clearly determined and fully understood by the valuer. The provisions regarding rebuilding and indemnity listed in the previous chapter will be clearly appreciated by a surveyor or architect, but there are insurance arrangements which should also have full attention in the valuation. By their nature these are insurance matters, but in many instances they affect the detailed work required in preparing a valuation for fire insurance purposes.

An example to illustrate an insurance provision in a policy is the application of a 'Day one' principle. There is more than one type of 'Day one' insurance, but the underlying basis of the cover is that the sum insured is notional and related to a certain date (usually the first day of the policy and called 'Day one'). The sum insured assumes that the building could be repaired, reinstated or reconstructed on that 'Day one' and allowances for inflation during the periods of demolition, site clearance, planning, reconstruction and the actual insurance period are all covered and provided by the wording of the policy and premium payments.

'Day one' valuations should be referred to as 'replacement' and not 'reinstatement' as the valuations will exclude those items necessary in addition to the notional replacement cost necessary to effect reinstatement.

Another illustration is the application of the 'first loss' principle. In simple terms a 'first loss' policy provides for insurance cover on any loss but with a sum insured related to one situation or one location. Where a valuation is required relating to a large risk spread over many properties and relating to a number of different locations, the insurance brokers may advise that a first loss policy should be arranged on certain specific items. Examples of this procedure could apply to sections like demolitions, professional fees and external items such as pavings, boundary walls, landscaping, fencing, gates, floodlighting, sewers and so forth. In these circumstances the valuation should advise on the situation that would provide the largest risk of a specified section and the first loss policy would have a liability related to the value of that particular situation. The largest risk situation for one first loss policy (e.g. demolitions) could be on one site and that for another (e.g. fees) could be at a different location.

Before proceeding to prepare a valuation for fire insurance purposes, the best advice is to call for a copy of the proposed policy and read it carefully. Pay particular attention to the basis of the sum insured and regulations with regard to payment against claims. Never be deterred from clarifying matters with the insurer and confirm any explanations in writing and record them in the body of the report and valuation.

Scope of reinstatement

A report and valuation on the sum insured should clearly state items and allowances which are included and those which are excluded. Apart from the general allowances of costs previously stated, the report should schedule separately each building included in the valuation, giving its plan reference, use, number of storeys, brief description of construction, and valuation costs as defined in the insurance policy. This not only defines the scope of the valuation but is extremely useful for reference and identification in the event of damage by fire. It is essential that plan references should relate to the fire plan prepared by the insurers. Where it is necessary for insurance purposes to know the estimated time required for reinstatement, this should be stated. Although a valuation report for fire insurance purposes may show separate figures for each plan reference, this would be used for rating the insurance premium only. The sum insured should be one figure for the whole location and based on a clear site with reinstatement in one continuous building operation.

Cost of foundations and drainage should be included in the calculations as most serious fires result in the need for their reconstruction. If the foundations are piled, the cost of piling can be excluded from the sum insured as it is unlikely to be affected by fire damage. Very deep foundations may also be excluded, but it is important to specify clearly such exclusions so that the costs are not used in average calculations, and the insured knows he is totally at risk on those items.

Services need careful consideration in determining whether they are regarded as insurable as part of the building or as plant. Decision of classification in this respect is between the insurer and the insured and often follows the manner in which items are regarded in the firm's accounts. Electrical supplies to machines may be listed as plant or building and likewise similar consideration should be given to lifts, hoists, sprinklers, smoke detectors, boilers, storage tanks, heaters, light fittings and many other service items. The important task is to list all relevant items, then to determine under which heading they are to be insured and to confirm in the report whether the items are included or excluded.

Similarly, a list of particular items should be considered which covers concrete pavings, tarmacadam, factory or estate roads, car and lorry parking, stacking areas, sewers and mains supply under factory roads, fencing, gates, ramps, dock levellers, road lighting, sewage pumps, greenhouses, garden sheds and all other external items. These items may not need insurance cover and it is most important to identify the component items of the premises and to indicate whether they are included or excluded from the valuation and consequently included or excluded from the sum insured.

In preparing a valuation recommending the sum insured for a building on a reinstatement basis for fire insurance purposes, the valuer must emphasise that his figures have a limited use. They should not be published without his written consent in case they are misinterpreted. The report should state that the valuation has no direct relation to market values nor can it be used in respect of current cost accounting for the basis of depreciated replacement value. Situations of fire insurance of large country mansions, whether they are used for their original purpose or in an alternative way, illustrate the need for the surveyor clearly to identify in his mind and his report that a valuation for replacement or reinstatement can be an entirely different figure from market value.

Planning requirements

Mention has already been made that current planning requirements may influence the surveyor's approach to preparing a valuation for insurance purposes. An existing factory constructed of steel and externally clad with corrugated iron could be permitted to remain in use indefinitely. The moment it is substantially damaged by fire the planning authority may require an entirely different type of building as replacement. The surveyor should make himself aware of the situation, explain the position to his client and provide in his report and valuation for either:

(a) A sum insured on an indemnity basis related to the market or book value of the factory which would become reclaimable from the insurers if the fire damage invoked a notice from the planning authority effectively requiring the damaged building to be totally removed, or;

(b) A sum insured on a reinstatement basis related to the cost of removing the remains of the existing damaged building and replacing it with one acceptable to the planning authority.

The insurance premium would be considerably less in (a) than in (b) for obvious reasons, but the security of the insured would be greatly increased in (b) since finance would be available to rebuild the factory. Additionally, in (a) no proper provision is being made for the situation of a small fire causing damage which would not provoke the planning authority to require total removal. The insured would then have to repair the damage and his indemnity insurance would prove insufficient due to average, for the reasons described in the previous chapter.

Where a Compulsory Purchase Order has been served in respect of a building and it has been agreed by the authority that demolition will take place on expiry of the Order, the insurance provision for the unexpired period is best arranged on indemnity. In the event of fire, the insured will have to admit that compensation in accordance with the Order will be received and that reinstatement is out of the question. In this situation the sum insured for the indemnity insurance should be related to the rental value of the building for the unexpired period of the Order.

A further consideration is the situation of a building being used in a different way from the general zoning laid down by the planning authority. A factory in an area planned for residential use provides an example. So long as the factory is not served with a Compulsory Purchase Order, the building would continue in use with existing user permission. In the event of serious fire damage, permission to rebuild may be refused. In these circumstances the insurance should be arranged for reinstatement on an alternative site and the sum insured related accordingly to the anticipated costs. In such a case several items normally excluded would now form part of the sum insured, including piling, concrete access roads, hardcore stacking areas, main services, sewers, general site layout items and even consideration of differing land values.

Listed buildings should have special consideration in determining the sum insured. The style of the building and materials to be used in any reinstatement may have very special cost implications. If the elevations have special qualities they may have to be preserved after a fire at considerable expense. Total demolition may not be permitted and the sum insured should reflect that possibility.

Domestic buildings

The sum insured for fire insurance of a domestic house or flat is dealt with by most people at some time in their lives. The principles of assessment are similar to those already stated, but the reasons for a marked discrepancy between purchase price and the proper sum insured should be understood.

The cost of construction of a single detached and isolated house will provide the basis for the calculations as previously described and this cost can be related to purchase price. Generally speaking, however, the construction of a house is part of an estate or group of buildings. In some instances the development of a very large area is determined and construction of similar types of units is undertaken over a relatively long period. The designer and builder have all the facilities to construct at very competitive prices. Bulk orders for materials can be placed, gangs of workmen can be organised in the most effective way and large items of plant can be used to great advantage. Most suburbs and new towns are developed in this manner and this mass production of units determines the purchase price of the house.

The fire insurance policy for a single house, however, is an individual contract between the insurer and the owner of the property. In the event of damage by fire the owner is faced with employing an architect and builder to deal with one house only, whether it be detached, semi-detached or one of a terrace. No longer is the architect concerned with hundreds of houses but just one, and no longer can the builder order materials in bulk and organise large gangs of workmen. The whole process of reconstruction becomes considerably more expensive.

Many domestic insurance policies do not invoke average, but in these situations the policy will probably include a clause allowing the insurers to repudiate the contract unless the sum insured is adequate for total reinstatement.

In dealing with a block of flats a consideration similar to the house situation arises. If the insurance contract is arranged separately flat by flat, very severe cost implications can develop. Fire damage to a penthouse on the top of a twenty-storey block of flats could mean the need for high cranes and expensive scaffolding in the reinstatement and all these eventualities would have to be reflected in the sum insured.

It is now common practice for the residents in blocks of flats to form themselves into a company. One of the responsibilities of such a company is to arrange fire insurance for the whole building. This allows for damage to one flat to be regarded as part only of the whole sum insured and the sum insured is related to reinstatement of the complete block and not to one flat. In the situation of a block of eight flats where the sum insured is £240,000 the expenditure in one flat of £40,000 for fire damage repair would be accepted by the insurers although only £30,000 would appear to be allocated to each flat.

Industrial estates

The industrial estate provides a similar situation to the housing estate. Experience has shown that the adequate sum insured on a separate industrial unit of a large estate should be approximately 50% higher than the proportionate cost of the original construction at the moment that the fire insurance responsibility is transferred from the contractor to the owner or occupier. Apart from the additional costs already mentioned in connection with housing estates, it is particularly important to appreciate that removal and reinstatement of a limited number of concrete or steel main frames in an industrial unit is a very expensive operation.

It is unusual and impracticable for users of an industrial estate to combine fire insurance as described for the flat complex. The grading of the use or production in each unit would affect the fire risk and the hazard and premiums would have to be considered separately. In a block of flats, however, the use and risk is constant throughout all units.

Commercial and industrial buildings

It is not possible to list all the complications which can arise in arranging the insurance of commercial and industrial buildings. In any event this is the responsibility of the insurance broker. However, the professional engaged on preparing the valuation for the sum insured should be aware of the more common situations that will influence a fire insurance valuation for these types of buildings.

- *Excess* – An insurance policy can be arranged so that the insured is entirely financially responsible for the first part of a loss. In this arrangement the amount which is uninsured is called 'excess'. Related to an industrial complex, the insured may decide to arrange the policy so that the first £10,000 of each and every claim is uninsured; an excess of £10,000. The benefit to the insured is a reduction in the insurance premium and the benefit to the insurers is a reduction in the risk. In these circumstances, the valuer should exclude from his valuation all separate buildings which cost less than £10,000 to reinstate and give the plan references of the buildings excluded and the reasons why. If the industrial complex is very substantial, like an oil refinery, the excess could be say £100,000 or even £250,000 and the understanding of this arrangement consequently becomes more pertinent to the valuer.

- *85% Rule* – Some fire insurance policies for commercial and industrial buildings contain a concession on the application of average. This concession provides that if the sum insured represents at least 85% of the total reinstatement cost of the whole, then average will not apply. In these circumstances, the insured is at risk for 15% of the reinstatement costs but it refers to the last 15% of the total reinstatement. In a complete destruction situation there would be under insurance, but it is exceptional for destruction to be 100% complete. Insurance brokers will quite rightly view this rule as a concession and not as a viable calculation to be taken into account in determining the sum insured. It was originally introduced to give some protection to the insured against inflation during the period of the insurance policy, which is invariably one year. When inflation is low, this provision can be regarded as an acceptable risk for this original purpose and can be considered particularly useful when insurance is arranged on an industrial complex when many separate buildings are covered under one accumulative sum insured.

- *Limit of liability* – There are many situations where commercial or industrial undertakings are responsible for insuring buildings or groups of buildings at separate locations. The options on how the fire insurance is arranged can be more flexible and instead of stating a sum insured, a limit of liability may be determined. This limit of liability would be considerably in excess of the maximum loss situation at any one location. It would still be necessary to determine the total value of the buildings at risk, but because of the vast spread of the risk the loss from any one incident would have a considerably reduced financial effect compared with the total. The calculation and determination of the insurance premium is a matter for the insurance broker and the insurers, but in these circumstances it is important that the valuer does not refer in his report to recommendations for a 'sum insured' because the risk will be covered by a stated limit of liability.

- *Obsolete buildings clause* – In the United Kingdom many industrial buildings are old and are used for a different purpose from their original design. The functional use of the building may be greatly reduced and in the event of serious fire damage reinstatement would never be considered. Where the building is at least fifty years old, the insured may elect to insure on what is called the 'obsolete buildings clause'. The sum insured is not related to the reinstatement cost of the existing building but to the construction costs of a new building on the same or an alternative site to suit present day requirements. When this clause is applied on a reinstatement basis its merit can be considerable. A heavily built engine shed constructed over fifty years ago, now being used as a general store, is a typical example which illustrates the situation. The existing building might cost £500,000 to reinstate, while a more suitable structure with less storey height and smaller floor area would cost only £200,000. By using the 'obsolete buildings clause' and a sum insured of £200,000 the insured would provide for a new building if the existing one was totally destroyed by fire. In the event of a small fire, where the need to repair the existing building arose, then the cost of repair up to £200,000 could be reclaimed without application of average. Once the repair costs exceed £200,000 then the whole of the obsolete building would be demolished and a new one built on the same or an alternative site.

Insurers introduced this clause to cope with the extreme, but all too common, situations of many of the buildings inherited and kept in use by the heavy and traditional industries. The insurance premiums are not reduced in direct proportion to the reduction in the sum insured, but a considerable saving of premium can be made by the use of this arrangement. The same principle can apply to insurance based on indemnity without provision for reinstatement. Many insurers now only offer the 'obsolete building clause' on an indemnity basis and in these circumstances the benefit to the insured is greatly reduced because depreciation has to be taken into account in calculating the indemnity compensation; full reinstatement reimbursement in this arrangement is not provided.

- *Long term agreement* – This refers to the period of contract between the insured and insurer. The policy issued is usually for one year but a discount on the insurance premium is allowed when renewal is guaranteed between the parties for a longer period of time. Long term agreements can be for any agreed period, but in the interests of both parties the term is usually restricted to three years. Sums insured still need to be reviewed and determined annually.

Consequential loss

In this chapter the subject throughout has been the material damage arising from the effect of fire on buildings. The same principles would apply to plant and machinery. There are two other widely known insurances, one being personal damage and the other consequential loss. Both these latter insurances are outside the scope of the architect and surveyor except in one respect. Consequential loss could be described as the 'knock-on' effect from material or personal damage. A simple example would be the effect of reductions in the profit of a commercial undertaking because the insured was deprived of the use of his building as a result of a fire. Included in those reductions of profit might be the need and cost to provide alternative accommodation while the building was reinstated. Additionally, if the building damaged by fire was leased, provision for loss of rent to the landlord may need to be considered; this latter point is often covered under the material loss policy of the building, by providing a sum equivalent to the rent multiplied by the estimated time to rebuild. Calculations for consequential loss can become very involved and complicated, but usually they are related in some way to an estimate of the total length of disturbance to the existing business resulting from a fire. The valuer therefore may be asked to advise on the maximum time necessary to rebuild in the event of fire, so that a view could be taken on the maximum time that profits would be affected. It is essential that the valuer eliminates all general consequential loss provisions from the valuation report and their effect on building costs.

Professional valuations

This chapter has shown the general approach required in determining the sum insured for buildings. Most prudent undertakings have professional valuations prepared and the advantages can be summarised as follows:

- The problems of a particular building or complex are reviewed in depth by an expert.

- The sum insured is accurate, giving adequate cover without waste of premium from overprovision.

- In the event of fire, the valuation will be generally accepted by the insurer as a sound basis for negotiation without application of average.

- The valuation provides an accurate basis for adjusting the sum insured during ensuing years.

- Measurements and details of construction are recorded in a proper manner and will be of great use for reference in preparing an insurance claim in the event of fire damage.

- Where variations in premium rates apply because of different types of construction or different occupancies of the buildings, the valuation of each building separately provides a basis of calculation of the fire premium.

It is good practice to have professional valuations reviewed annually. After the initial inspection and valuation the next two or three years can be adjusted by reference to building cost indices and each three or four years adjusted by a site inspection and detailed reconsideration. Regular inspection of large factory complexes ensures that the inevitable changes that take place in buildings are properly adjusted in the sum insured.

In fire insurance valuations, large sums of money are usually involved and the professional adviser is himself at some risk. He should notify his own professional indemnity insurers of the extent of his work in this field and ensure that his insurance cover is adequate.

Example of valuation

It is impossible to provide an example of a professional valuation for fire insurance purposes to illustrate all possible situations. The following example is typical for a small industrial complex where alternative valuations are given to enable the insurance brokers and the insured to consider the comparative cost of using a traditional reinstatement insurance policy or using 'Day one' cover.

VALUATIONS FOR BUILDINGS
– for –
FIRE INSURANCE PURPOSES
– for –
ENDURABLE LAMPS LIMITED
– in respect of their premises at –
LOW LANE, HORSFORTH, YORKSHIRE

The following figures are the valuation costs of reinstating the buildings after fire damage, with materials and in a style similar to that which now exists, including the cost of removal of debris and all necessary shoring and protection to existing and adjoining buildings.

We have included against each plan reference the cost of the following items where applicable:

- Superstructure and substructure including ground floor slab and finish.
- Services below ground floor slab.
- Plumbing and electrical installations.
- Sprinkler installation.
- Air handling installations.
- Heating installations.
- Boilers and oil storage restricted to heating and hot water installations.
- Lifts.
- Name signs.
- Light fittings.
- Demolition and site clearance.
- Professional fees.
- Local authority fees.

No allowance has been made in the valuations for the following:

- External pavings, roads and open car park areas, with the exception of tarmacadam paving where adjacent to buildings.

- Main services under external pavings.

- Underground storage tanks.

- Boiler for process steam and pipe lines.

- Compressed air lines.

- Racking in stores.

- Canteen and kitchen equipment.

- Carpets and other fitted floor coverings.

- Demountable partitions.

- Furniture and plant generally.

- Sub-station equipment.

The valuation makes no allowance for VAT. Charges will be made in this respect to a building contractor for materials and services, but he should be able to recover such tax from HM Customs and Excise as laid down in the Finance Act 1972. You should consult your insurance broker on the matter of any VAT charge arising between a builder and yourselves in the event of reconstruction or partial reconstruction after fire damage. If however you are VAT registered, in our view no additional allowance to our valuations needs to be made in this respect.

The basis of the valuations under columns 'A' is building prices current at the 31 December 1983. The figures given are the notional replacement costs on that day ('Day one') and *exclude* all provisions in respect of inflation relating to:

Increased costs for the period of planning, tender and site clearance.

Increased costs for the period of reconstruction.

Increased costs for the period of insurance.

In respect of the column 'A' figures, it is essential to confirm that adequate provision for the above exclusions will be covered by the wording of the policy.

The basis of the valuations under column 'B' is building prices current at the 31 December 1983 but including provisions in respect of inflation relating to:

Increased costs for the period of planning, tender and site clearance.

Increased costs for the period of reconstruction.

In respect of the column 'B' figures, the escalator addition relating to the period of insurance should be 8% per annum.

Your insurance broker will also be able to explain the working of the 85% rule usually applied in respect of industrial and commercial properties without application of average and its possible effect on the sum to be insured.

No investigations have been made by us that planning approval would or would not be given in the event of serious damage by fire to any of the properties in this report.

The valuation figures do not allow for any improvements in the reconstruction of existing buildings in the event of reinstatement after fire damage with the exception of the minimum requirements to comply with building regulations. Improvements to comply with planning requirements are excluded from this valuation.

The valuation figures given for each plan reference are provided to assist in determining the fire rating and premium for that particular building. The sum insured should be based on a total loss situation for the whole location. Any partial loss, whether it be a complete building or section of a building, is to be regarded as a part of the whole sum insured for this location and your insurance policy should be clear on this matter.

The valuation figures are based on the assumption that in the event of fire damage, adequate time would be available in order to plan the reconstruction in an economical manner and that competitive tenders would be obtained from builders. The effect of any attempt to shorten the reinstatement period to reduce consequential loss claim is excluded from these valuations.

The existing automatic sprinkler installation does not appear to comply in all respects with the requirements of the current Edition of the F.O.C. Rules on this subject. Our valuation allows an additional sum of £10,000 in Plan Ref. 1 and 2. (column 'A') and £10,900 (column 'B') for the cost in reinstatement to comply with these Rules. You are advised to have incorporated in the insurance policy a special memorandum to cover the additional cost that would be incurred in any reinstatement after fire.

The valuation must *not* be taken as a guide to the current market values of the properties and reference to these valuations may not be included in any published document without the surveyor's written approval.

Plan Ref. 1 and 2		'A'	'B'
Factory, warehouse, workshops, boiler room, laboratories, toilets, canteen, kitchen and air handling plant areas. Two storey and plant mezzanine over part of area. Brick and PVC covered steel wall cladding; reinforced concrete and steel framing; concrete floors; flat felted and decked roof; air handling and heating installation; sprinklers including storage tank.	))))))))))))	3,250,000	3,620,000
Plan Ref. 3 Offices. Three- and four-storey and roof housing. Brick walls with slate dressings; reinforced concrete framing, floors, staircase and roof; good quality internal and external finishings; brick and concrete external fire escape stairs.	))))))))	840,000	930,000
Plan Ref. 4 Sprinkler pump housing. One-storey. Brick walls; concrete floors; concrete roof.	))))	15,000	16,500
External Items Brick retaining walls. Railings. Suspended concrete terrace. Tarmacadam paving. Shed.	)))))	56,000	61,500
Totals		£4,161,000	£4,628,000

'Smoke damage'

Chapter 10

AFTER THE FIRE

The professionals involved

When a building has suffered damage by fire, certain clauses of the insurance policy may begin to operate. If the damage is very limited and simple, the matter may be resolved directly between the insured and the insurers. However, in many cases of fire the eventual outcome cannot be settled without involving professional advisors to act for both parties. In order to perform in a proper and professional manner it is essential to understand the scope and responsibility of those involved.

The first responsibility must be with the insured who has to notify the insurer that a claim arises; probably first notified by telephone and later confirmed by completing a claim form. It is of paramount importance to appreciate the fundamental difference between the objectives of the insurer and the insured. The insurer will be dealing solely with financial compensation arising under the terms of the policy; the insured will not only be seeking that compensation but in many cases will also be seeking to reinstate the building from the funds obtained.

The insurer has a commercial responsibility to keep the amount of compensation to the minimum commensurate with his legal obligation and business probity. He will almost certainly appoint a loss adjusting firm to act for him and, if the claim hinges on a legal matter, a solicitor to protect his interest, acting in close liaison with the appointed loss adjuster. These professional parties will be appointed direct by the insurer.

On the other hand, the insured is unlikely to be experienced in claims or in reinstatement after fire and will often not know to whom to turn for proper advice or the co-ordination of his requirments. Without a co-ordinator or team leader to direct his activities, the insured can find himself poorly represented and at a disadvantage against a highly experienced professional team acting for the insurer. The architect and quantity surveyor have the qualifications to handle this position, but either or both are often appointed too late or appointed with limited instructions. In addition, insurance matters are on the periphery of their expertise and therefore instead of taking charge they tend to await instructions from others. In fact they should be discharging a duty well befitting their professional capabilities.

Chapter 11 will discuss the claim, but before dealing with this matter the duties of those involved after the fire should be reviewed. In broad terms they can be summarised as follows:

- Insurer (dealing with a financial claim):–
  Insurer – Loss adjuster – Architect/Engineer
                               – Quantity surveyor
                               – Accountant
        – Solicitor

- Insured (dealing with reinstatement and compensation or a financial claim):
  Insured – Insurance Broker
           – Assessor
           – Architect/Engineer
           – Quantity Surveyor
           – Accountant
           – Banker
           – Landlord
           – Solicitor

The insurer

The insurer is the person taking the risk of damage to the material asset for which he receives a fire insurance premium. On notification of a claim he will ensure that:

- The policy is currently in force and that all premiums due have been paid.

- The details have been issued clearly defining the conditions of the policy.

- The damage applies to the property insured.

- Liability for the damage is covered by the policy.

- Third parties (e.g. landlords) named in the policy are notified.

- A loss adjusting firm is appointed and is provided with all the details of the policy.

The insurer will deal with recommendations from the loss adjuster including action relating to interim and final decisions and payments. If liability for the claim is denied or limited, the insurer should inform the insured as quickly as possible.

The loss adjuster

The loss adjuster is qualified by professional examination through the Chartered Institute of Loss Adjusters. He is appointed by the insurer and his duty is primarily to approve or adjust a financial claim submitted by the insured so that he can recommend payment by the insurer.

It is usual for a loss adjuster to receive his instructions only from insurers and the majority have no other type of client. Much of his work would be of a general nature, but larger firms employ adjusters who specialise in particular subjects. A loss adjuster qualified in accountancy would be well suited to deal with consequential loss claims relating to reduction of profits. Likewise, a quantity surveyor working as a loss adjuster would be best suited to deal with a large fire insurance claim on a building, while someone qualified in law could best handle professional indemnity claims. Many adjusting firms do not employ professionally qualified people to handle all types of loss and therefore rely on other professionals to assist them from time to time. In these situations it is usual for the independent professional to receive instructions and payment of fees from the adjuster.

Because the adjuster's prime object is to limit the liability of his client in accordance with the details of the insurance policy, his duties can be far-ranging. At times he may appear to assume responsibilities beyond the professional duties to his client, but in fire damage situations he is experienced and his intentions will always be to indemnify the insured so that he is put back into a position equal to that before the damage. The loss adjuster will therefore be concerned with the following duties when called to an incident of fire damage to a building:

- Investigate the cause of the fire and confirm that the insurer is liable and that the policy should come into operation.

- Obtain and study the fire brigade report of the incident and consider the influence it may have on the operation of the policy.

- Ensure that photographs are taken of the damage immediately after the fire.

- Ensure that the maximum is obtained from salvage. This applies particularly to plant and machinery which can have a high resale value overseas.

- Approve the method and cost of first stage demolition, temporary protection and support of buildings owned by the insured and third parties.

- Establish whether the claim will be on an indemnity or reinstatement basis.

- Approve the extent of second stage demolition. In this situation an independent architect or engineer may be instructed by the adjuster to give opinions on the extent that structural work needs to be removed and renewed.

- Approve the specification of the work to be carried out to reinstate the building following the fire damage. In certain situations, as described in previous chapters, this could involve the construction of a new building comprising different types of materials, different floor and storey heights and even constructed at a different location.

- Approve the method of obtaining tenders or determining the building costs. The advice of an independent quantity surveyor may be of considerable value on this matter.

- Offer the facilities of discount arrangements on bulk items. Many adjusters have this arrangement with suppliers of carpets, office equipment and similar items.

- Consider the effect of the reinstatement programme on the consequential loss claim and ensure that the overall solution limits the overall liability of the insurer to the minimum. The services of an independent accountant could be essential in advising on the correct balance between reducing reinstatement costs through competitive tenders based on economic reconstruction time and loss of profits from the business.

- Consider the requests and need for providing alternative accommodation for the insured so that business can continue with the minimum of disturbance. In the case of domestic property a recommendation may be required on the alternative costs of renting temporary accommodation or reimbursing hotel expenses.

- Consider the adequacy of the sum insured and the possibility and effect of applying average.

- Consider excess provisions in the policy.

- Consider and recommend interim payments.

- Correlate the claim for building damage with other insurance claims arising from the same incident and liaise with other appointed loss adjusters.

- Consider and recommend the financial settlement.

As a result of extensive damage by fire, several policies may come into operation and these policies may be with different insurers. It is possible therefore, and not uncommon, for a fire to invoke separate policies from different insurers on the buildings, plant, machinery and contents and the consequential loss of profit in a business and for each insurer to appoint a different loss adjuster. There are rules of etiquette in this matter and the insured is not usually placed in a position of disadvantage. However, there can be advantages to the insured in placing his insurance on building, contents and consequential loss with one insurer, so that in the event of a claim for fire, one loss adjuster would be appointed.

The need to understand the business activities of some larger industrial firms has persuaded insurers to appoint loss adjusters on a long term basis. This ensures that when a loss occurs, the adjuster will be on site within hours and dealing with persons well known to him over a period of years. In addition, he will understand the working of the business and be in a position to assist in approving proposals to minimise the loss.

The insured

The insured is a client of the insurance industry; he pays the insurance premiums and is relieved of the risk. If fire damage is strictly limited, it may be rectified in a simple manner and both parties may agree that competitive prices from two or three builders will be the basis of the reinstatement and the claim; an example would be smoke damage to a room with redecoration providing the reinstatement. Once the damage is no longer simple, the insurer will be wise to employ consultants. In all cases the insured or his consultants should ensure that:

- Proper written notification of the incident is made to the insurers.

- Notification and details of the appointment of other consultants is given to the insurance broker involved.

- The landlord is informed where the insured is a tenant.

- The full extent of the damage is assessed and the possible effect on the running of a business understood. This should lead the insured to appointing the consultants best suited to limit his loss and to reinstate his building and his business as quickly as possible and at the most economical cost. Consultation with the appointed loss adjuster is essential on these matters.

- Prompt attention is given to finding suitable alternative accommodation for business or domestic use where the damage warrants this requirement.

The insurance broker

The principal duty of an insurance broker is to place his clients' risks with insurers at the most favourable terms.

Where a claim on a policy arises, the broker will ensure that the interests of his client are protected. Although some brokers have 'claims' departments, where the loss is serious fire damage, they usually recommend the appointment of independent professional firms to handle the claim and the reinstatement.

Assessor

An assessor schedules and costs the damage and presents a claim on behalf of the insured to the loss adjuster. In the majority of cases his duty is restricted to negotiating the financial compensation while the responsibility of reinstatement remains with the insured. The assessor could be regarded as the direct counterpart to the loss adjuster in that both would have responsibility for determining financial matters on insurance but usually neither is directly involved with reinstatement.

Architect and engineer

The insured will appoint an architect or engineer to take the responsibility for protecting his building and arranging for reconstruction after the fire damage. It is essential that this architect's or engineer's brief is clearly defined at an early stage. In particular, he should have a complete understanding of the financial implications of the likely compensation from the policy and the cost of reinstatement or reconstruction. From the beginning he should liaise closely with the loss adjuster and reach written agreement step for step. His duties therefore are likely to be to:

- Determine the extent of first stage demolition of any dangerous structure.

- Arrange first-aid repairs and protection of the undamaged sections of the building.

- Arrange temporary support of adjoining buildings.

- Ensure the maximum is obtained from salvage.

- Have extensive photographs taken at each stage of demolition, protection and temporary support and obtain the written agreement of the loss adjuster that they represent a true record of the damage and extent of work required to be rectified.

- Determine the brief with the client and, assuming reinstatement is to take place, proceed with the following:

Arrange for second stage demolition or total site clearance.

Prepare specifications and drawings for the reinstatement and obtain all necessary consents and approvals.

Determine with the loss adjuster the basis and programme of the building contract and obtain tenders or negotiate a contract.

Supervise the reinstatement.

Quantity surveyor

The duties of the quantity surveyor will be determined by whether or not the material damage to the building is to be reinstated. If reinstatement is not to take place, the quantity surveyor will have duties identical to those of an assessor. He will prepare a financial claim for negotiation and agreement with the loss adjuster.

If reinstatement is to take place, the quantity surveyor will act in the normal way in conjuction with the architect in preparing documents to obtain tenders and subsequently assist with the running of the contract. The cost evidence and information from the building reinstatement contract may be sufficient to prepare the insurance claim and get it agreed with the loss adjuster. If this is not the case then a separate document may be required defining the extent and cost of the damage recoverable under the policy.

The quantity surveyor should consider the following:

- The financial implications of insurance policy and correct understanding of the clauses. Confirm whether the policy is based on indemnity or reinstatement.

- The effect of average.

- The effect of deductable excess.

- The need to prepare a priced Schedule of Loss.

- The programme of reinstatement work, relating it to any consequential loss situation.

- The effect of VAT on builders' accounts and whether it is recoverable under the wording of the policy.

- The whole financial programme, agreeing it step for step in writing with the loss adjuster before committing the client to expense. In the case of reinstatement, the proper arrangements for interim payments from insurers.

- The need to define carefully the duties of each consultant and establish who will be responsible for paying respective fees where more than one party is insured.

Advisers and interested parties

Although the insured is primarily responsible for initiating the claim and reinstatement, it must be accepted that, having suffered the loss of his business or home, his actions may not be rational. Whoever therefore takes charge in the first day or two should keep an open and clear mind on the necessary action to be taken. The period immediately after the fire is crucial.

Landlords, bankers, mortgagors must all be advised and their interests disclosed. If the damage is to a manufacturing business then customers and suppliers must be advised. Loss of contents and stock will need to be scheduled and probable consequential damage to business profits will need to be reviewed. In a complicated incident the services of an accountant will be essential. His duty will be to examine the financial effect of each course of action recommended so that a rational overall solution is determined. Architects and quantity surveyors tend to consider building costs and programme as paramount to all other factors. In many instances the financial loss due to waste of stock or decline in business can be the decisive factor in determining the correct and most economic remedy.

The solicitor

Where a dispute on liability or interpretation of the policy cannot be settled between the insured and the insurer, or their representatives, the matter may then become a legal action. These circumstances are unusual and case law on fire matters is limited; where a dispute is referred to lawyers there is a general desire by the insurer to settle out of court. Nevertheless, points of principle have come before the courts and some case law now exists.

Action through law can often be protracted and is likely to concern liability and quantum. The responsibility for the actual reinstatement (if at issue) will always rest with the insured and he should appreciate that recourse to lawyers can often delay the financial settlement of an insurance claim and therefore deny him the advantage of availability of cash to rebuild. Where the matter of a claim for fire damage to a building is put to lawyers to resolve, parties should bear in mind that:

- The solicitors appointed should be knowledgeable and experienced in insurance and building matters.

- Commercial negotiations may still continue independent of the legal action.

146

- The parties to the dispute should inform the solicitors of the circumstances in clear terms so that the lawyers can determine the points at issue. Litigants should avoid attempting to decide on legal issues.

- Solicitors should determine the best practical approach to protect the client's interest. This usually means, in the case of the plaintiff, commencing proceedings as quickly as possible so that the points at issue are properly defined, thus crystallising the parties' attitudes towards settlement. The defendant (inevitably the insurer) will equally want the matter defined in legal terms at the earliest opportunity.

- Considerable legal costs in these cases are incurred before trial. In broad terms, a successful litigant will recover approximately only 60% of his legal costs.

- Contributions to or from third parties must be clearly defined.

'Credibility of information'

Chapter 11

THE CLAIM AND THE REINSTATEMENT

Basis of claim

The responsibility for preparing and presenting the financial claim under any insurance policy rests entirely with the insured. The need to determine the basis of the claim is paramount at an early stage. The alternatives are:

- Reinstatement costs
- Indemnity cash payment.

Reinstatement costs

Where the insurance policy contains a reinstatement memorandum, the insured has the opportunity of recouping the cost of putting back the building with materials and in a style similar to that which existed before fire damage – in simple terms, 'new for old'. This will apply to all aspects of the building, including items having a limited life such as decorations and service installations.

Reinstatement has a special meaning when used in insurance. In literal terms, reinstatement means to re-establish in former position and to restore in proper order. Insurers usually consider reinstatement in this strict manner. Architects and quantity surveyors often confuse reinstatement with reconstruction. Where the reconstruction ensures that the style, size and performance are similar to before fire damage, the insurers will regard this as reinstatement. However, where damage is extensive, there is the opportunity of reconstructing in a different way. Perhaps reinforced concrete may be used in lieu of timber construction or the size of the building may be altered. Then reconstruction is not pure reinstatement but contains some degree of betterment and in this situation a financial reduction will have to be made from the reconstruction costs. Where these alterations are made to comply with building regulations or planning requirements they are accepted as reinstatement providing the policy contains the 'Local Authority' clause. But if the changes are a choice by the insured, then the whole reconstruction is not regarded as reinstatement.

Once reinstatement is established and agreed by the loss adjuster, the documentation follows the course of normal building work. The designer prepares drawings and specification notes describing the extent of the work and these are agreed with the loss adjuster. The quantity surveyor prepares bills of quantities for competitive or negotiated tender and the interim certification proceeds as normal.

The contract to reinstate will be between the insured and the building contractor. Likewise the engagement of professional consultants will lie directly with the insured. Payment by the insurers is regulated by expenditure by the insured. This applies to fees as well as to building costs. It is usual to arrange with the loss adjuster for him to recommend interim payments from the insurer which will follow closely on architect's certificates. These payments will normally be from the insurer to the insured but there may be cases when the insurer will decide to pay the building contractor direct. The total expenditure on the work will form the claim and in such a case the insured should be reimbursed his total outlay.

Where a building is reinstated it is unusual for payment to be made by the insurer in advance of work being carried out. However, there is a point of view that, as all insurance policies are indemnity, the full indemnity, when determined, should be paid in advance of reinstatement. The balance arising from the reinstatement memorandum over the indemnity value would then become due if the cost of reinstatement exceeds the indemnity. Although many adjusters resist this view, it is a matter of negotiation and underlines the real necessity to agree fact and procedure with other professionals at an early stage and in a regular way.

It is worthy of repetition to state that reinstatement can be accepted on an alternative site and in a modern design if planning requirements and Building Regulations are to be properly met. Once the principle has been agreed with the loss adjuster the procedure will be similar to that outlined above.

Indemnity cash payment

The indemnity cash settlement provides for a lump sum payment as compensation under the terms of the policy. Once the indemnity sum is agreed it is paid in full by the insurer and his obligation is fulfilled. How the money is used and what happens to the damaged building is no longer any concern of the insurer. The basis of this compensation can be derived from:

- A priced schedule of loss.
- Reconstruction costs.
- Market values.

Priced schedule of loss

A priced schedule of loss represents the notional reinstatement cost of fire damage to a building. The schedule can be prepared as a specification or as a bill of quantities. It will contain all the items necessary to reinstate the building exactly as it existed before the damage. It is not relevant whether the materials described would be used in the reconstruction. The schedule might therefore contain items of lead plumbing, king post roof trusses, hardwood beams and joinery, timber lath and plaster and many other items no longer used in modern day construction. The information for the schedule would have to be drawn from the items of debris, past drawings and photographs if available, and from question and answer with the insured.

The document will be unsuitable to obtain tenders and will need to be priced in a notional manner. The total cost of the schedule will represent the gross reinstatement cost of the damage using materials and in a style similar to existing. If the policy is one of indemnity the pricing will be related to the date of fire. Should the policy contain a reinstatement memorandum then the basis of pricing will include allowance for inflation during the period of planning and reinstatement.

Once this document has been checked and agreed by the loss adjuster, the question of depreciation (or betterment or dilapidation) will need to be considered. This is the point where the reinstatement cost is converted into indemnity value. Loss adjusters generally have a very realistic view of this matter. Bricks, stone, concrete, slates, steel, glass and many other materials used in a building seldom lose their qualities unless the whole building is allowed to decay. Electric wiring, heating installations and decorations do depreciate with age and it is usually to these items that a good negotiator can restrict the adjuster on dilapidations. The adjuster may consider each item for depreciation but it is usual for negotiations to centre around a percentage figure to be applied to the whole or sections of the priced schedule.

It is not possible to obtain a cash payment from an insurer equivalent to the reinstatement cost if the reinstatement is not carried out, even though the policy contains all the right clauses. The cash payment will always be reduced to the indemnity basis of valuation.

There are two legal cases which support this method of settlement, namely *Reynold and Another* v *Phoenix Assurance Company Limited* (1978) and *Pleasurama Limited* v *Sun Alliance Insurance Company Limited* (1979). Mention is made of these cases as unsuccessful attempts have been made by insurers to set aside indemnity cash payments related to notional reinstatement costs depreciated for either wear and tear or other factors.

Reconstruction costs

In order for reconstruction costs to be used in support of an indemnity cash payment a flexible and understanding approach by the loss adjuster is required. There can be many instances of this type of negotiation and two brief examples can illustrate the possible procedure:

- Following fire damage the insured decides to reconstruct to provide a larger building but retaining the similar style to existing. By relating floor areas, the reconstruction costs of the whole may provide the reinstatement cost for the section damaged. Depreciation would then apply as before.

- Fire destroys two bays of a complicated industrial building and the insured decides to reconstruct in a simple manner. The reconstruction costs, together with a priced schedule of loss for the complicated sections not reinstated, may then provide the necessary information to agree an indemnity payment for the loss.

Very careful thought should be given to the best method of resolving the claim and the credibility of the information provided is important. Reconstruction costs properly obtained in competition provide a sound basis for use in negotiations.

Market values

If a loss can be related to market values then indemnity may prevail and a reinstatement memorandum may be discounted. A legal base for this point of view is provided in the case of *Leppard* v *Excess Insurance Company Limited* (1977). The case shows a cottage totally destroyed by fire and insured with a reinstatement memorandum in the sum of £14 000. The cottage had not been reinstated. The cost of reinstatement after taking betterment into account was estimated at £8694. There was admitted evidence that the insured had tried to sell the cottage for a considerable period before the fire at a figure of £4500. The land was valued at £1500. All the figures were agreed by the parties prior to the hearing. The Appeal Court ruled that the correct damage was £3000, being the difference of the market value before fire (£4500) and the market value of the land (£1500). The case shows that with these circumstances the damages (or claim) relate to the indemnity arising from the insurance contract; it was not an action (or claim) for specific performance which effectively arises from the application of a reinstatement memorandum.

Insurers generally respect the adage 'an Englishman's home is his castle'. This not only relates to domestic but industrial and commercial properties. If a policy contains a reinstatement memorandum, then in most instances the insurer will agree specific performance of reinstatement and refund the costs. Should there be evidence that the insured has previously sought to sell his property, or it can be demonstrated that similar is available without detriment to him, then insurers may look very closely at this situation and consider market values.

Partial loss

The majority of fire damage to buildings represents a partial loss. The principles of reinstatement and indemnity similarly apply. Situations can show advantages by using both. An illustration would be an old-fashioned building now used as a printing works where a serious fire destroys the first floor and roof but leaves the ground floor in a repairable state. Agreement can be reached with the loss adjuster for the ground floor to be repaired on a reinstatement basis including providing a suitable roof in lieu of the first floor construction. The upper floor and roof might then not be rebuilt as the owner can find better use for the cash compensation. The repair cost of the ground floor storey would be reimbursed in full on a reinstatement basis and the notional reinstatement cost of the first floor storey reduced for dilapidations would form an indemnity cash settlement.

Professional fees

Professional fees incurred in connection with reinstatement are usually reclaimable from insurers, provided the sum insured is adequate and the policy contains provision to pay fees.

Professional fees incurred for preparing the claim are not reclaimable through an insurance policy. The cost of preparing a priced schedule of loss would not be admitted in a claim. Indemnity cash settlements do not generally include allowance for professional fees. However, examples are given in this chapter to show how reconstruction costs can be used as a basis for indemnity settlement and in these cases the adjuster could recommend reimbursement of fees.

Average

A claim will be subject to the adequacy of the sum insured. Average has been explained in chapter 9, but it is in the claim situation that it is applied both in respect of reinstatement and indemnity.

Excess

In a like manner, excess is taken into account when finalising a claim. Generally this is a simple deduction following the agreement of the details of the claim. Complications can arise if the deductable excess relates to 'each and every incident'. In some situations, particularly in cases of malicious damage like arson, there can be several 'seats of fire' and insurers may take the view that each 'seat of fire' is a separate incident. Circumstances will determine how insurers view such a case and usually they are not severe in their approach. However, bad advice can be given by professionals if all the facts are not properly assembled and fully understood.

Long term liability

Redundant parts of a building may be damaged by fire but require no immediate attention in repair. An example would be damage to an original plastered ceiling now hidden by a suspended false ceiling. Where the building is leased, the original ceiling may be regarded as part of the structure and insured by the landlord; the suspended ceiling could be the responsibility of and insured by the tenant. At the end of the lease, the suspended ceiling may be removed and the damaged plaster ceiling would then be exposed. The landlord would need to protect himself against this deferred situation and this is usually achieved by exchange of correspondence and provision by the insurers of a sum set aside for the defined period. The sum would need to be properly invested to provide an allowance for inflation in building costs.

Fire during construction

If fire occurs during new construction or refurbishment, the claim for loss will be the responsibility of the insured in the normal manner. If the terms of the building contract require the contractor to insure then he will make the claim, whereas it will be the employer if he is the insured.

Generally the terms of building contracts define the provisions for insurance and the possible events following fire damage in respect of reinstatement. For example, in clause 22A of the JCT Standard Form 1980 Edition, provision is made for insurance to be in the joint names of the employer and the contractor. The contractor is responsible for arranging the policy and paying the premium and he is instructed to insure for the full reinstatement value of the building together with the cost of professional fees. No specific reference is made to provision for inflation costs during the reinstatement period and if the sum insured is the same as the contract sum this could prove inadequate and any claim may be subject to average.

The above JCT clause details the procedure for repair and defines the method of payment to the contractor for the reinstatement costs. Sub-clause 4.2 of this clause implies that the architect has some responsibility to the insurer, but this could only apply if it is written into the policy. What is not stated in this clause, and is omitted from most building contracts, is insurance provision relating to loss or expense by the employer being denied the use of the completed building as a result of fire damage. This could be loss of rent, presumably equivalent to the amount of the stated liquidated damages – a consequential loss situation. If proper insurance has not been provided to cover this point, no claim for compensation could be made on the building contract policy.

The compensation to the contractor may also not be completely provided by solely complying with insurance requirements in a building contract. He too may have a consequential loss situation in that the building contract may not provide him with adequate protection to meet any loss or expense for extended contract period arising from delay by fire damage.

This matter has been dealt with under one heading. The need for adequate insurance arises at the time of placing the risk and, as in all insurance, the claim will relate to the wording of the policy and the sum insured. Employers and contractors should carefully review their insurance cover prior to proceeding with a building contract and consider the effect fire would have on their affairs if damage occurred during construction.

Value Added Tax

Value Added Tax arises from the Finance Act 1972 and is subject to amendments thereto.

The Finance Act 1984 appears to provide that if a building is reduced by fire to ground level, or if no more than a shell remains, the work of reinstatement is zero-rated (see also VAT Notes (No. 1) 1983/4). If more than a shell remains, the repair work to reinstate would be standard-rated. Previously, elements of alteration work separately identified in reconstruction after fire were zero-rated but this provision ceased with the Finance Act 1984.

If the insured is registered with HM Customs and Excise he should be able to reclaim from them any VAT properly chargeable by the builder relating to fire repair. In these cases the insurers will not admit VAT and it is not recognised in the claim.

Where the insured is not VAT registered, the claim should include the correct percentage addition and this should be met by the insurers. Professional advisers should ascertain their clients position in this matter, as a claim arising from partial loss is likely to carry a substantial amount relating to VAT.

There are anomalies in the application of VAT in dealing with charitable organisations. These should be carefully researched to ensure that proper professional advice is given.

Repair

Whether the fire damage repair is reinstatement or reconstruction in a manner different from the original, various alternative tendering procedures and contractual arrangements are available. These are dealt with in detail in The Aqua Group book entitled *Tenders and Contracts for Building*.

The professional team should select the best combination to suit the particular circumstances. As an *aide memoire* to this subject the list below gives the principal options available.

Tendering procedures:

- Competition.
- Negotiation with one builder.

Tender Documents leading to Contract Documents:

- Bills of Quantities.
- Bills of Approximate Quantities.
- Schedule of Rates.
- Specification and Drawings.
- Cost Reimbursement.
- Target Cost.

Contractual Arrangements:

- Firm Price.
- Fluctuating Price.
- Standard Forms with or without Quantities or with Approximate Quantities.
- Fixed Fee Standard Form of Prime Cost Contract.
- Standard Form of Agreement for Minor Building Works.
- Management Contract.

Appendix 1
LEGISLATION FOR ENGLAND AND WALES

General legislation – England and Wales

- Building Regulations 1976 (See chapter 3)

  Part E concerns safety in fire; section 1 – Structural Fire Precautions and section 2 – Means of Escape in Case of Fire.

- The Health and Safety at Work (etc.) Act 1974

  Section 78 concerns amendments to the Fire Precautions Act of 1971 which, by reference to places of work, affects the Factories Act 1961 and the Offices, Shops and Railway Premises Act 1963. Also reference is made to fire related matters such as Highly Flammable Liquids and LPG Regulations 1972.

- The Fire Precautions Act 1971

  An enabling Act concerning means of escape and fire precautions in existing buildings in use. It provides that a Fire Certificate shall be issued for certain types of buildings as shall be designated from time to time by order of the Secretary of State. Designating orders so far made are the Fire Precautions (Hotels and Boarding Houses) Order 1972; also The Fire Precautions Act 1971 (Modifications) Regulations 1976, The Fire Precautions (Factories, Offices, Shops and Railway Premises) Order 1976 and (Non-Certificated Factories, Offices, Shops and Railway Premises) Regulations 1976. These latter orders incorporate into the Fire Precautions Act provisions which were previously part of the Factories Act 1961 and Offices, Shops and Railway Premises Act 1963.

- Public Health Act 1936 amended by the 1961 Act

  Enabling legislation for the Building Regulations; section 59 concerns escape from places of public assembly; section 60,

159

escape from sleeping accommodation above the first floor or in schools 20 ft above ground level; children's homes and lettable dwellings. Act administered by the local authority or fire authority.

Local Acts – England and Wales

A variety of local authority Acts including references to fire fighting access, fire precautions and means of escape in large or tall buildings, warehouses, etc. These are subject to amendment in the light of pending legislation and designers must make a point of consulting local officers.

Fire related legislation – England and Wales

- The Offices, Shops and Railway Premises Act 1963

  Contains fire precautions which have been transferred to the Fire Precautions Act, now applicable only to buildings having existing OSRPA Certificates; refers to separate tenancies rather than whole buildings.

- The Factories Act 1961 (Amended by Statutory Instruments 1976)

  General fire precautions are now transferred to the Fire Precautions Act except for buildings involving process hazards.

- The Housing Acts 1961 and 1969

  Section 16 concerns the provisions considered necessary by local authorities in lodgings and housing multi-occupancy. Section 60 empowers local authorities to make closing orders on such premises.

- Education Act 1944

  General standards of construction and safety required by Department of Education and Science for non-maintained schools.

- Childrens and Young Persons Act 1969 Childrens Act 1948 Nurseries and Child Minders Regulations Act 1948 Health Service and Public Health Act 1968

  Fire precautions and registration of community and voluntary homes for children, nurseries, etc.; Acts administered by the local authorities.

- The National Assistance Act 1948
The Mental Health Act 1959
Nursing Homes Act 1975

Concerning the registration of old people's homes and private nursing homes

- The Cinematographic Acts 1909 and 1952 and Regulations Made Thereunder
The Private Places of Entertainment (Licensing) Act 1967
The Theatres Act 1968
The Sunday Theatres Act 1972

These Acts refer to places of assembly, entertainment and exhibition which require licence for their use by the local authority. Licence will not be granted without conformity to regulations concerning means of escape and other relevant matters. Premises for cinematograph exhibitions must comply with the Cinematograph (Safety) Regulations of 1955, 1958 and 1965.

- The Chronically Sick and Disabled Persons Act 1970

The Act concerns special provisions for parking, access, escape, etc., which it is reasonable to make for disabled people using or visiting any premises; all in addition to the requirements of the Acts listed above. Section 8 refers to similar provisions in schools and universities.

- The Licensing Act 1964

Refers to premises such as restaurants and clubs where liquor may be supplied or sold under licence, such licence being withheld should the fire authority be dissatisfied with the provision of means of escape and other fire safety measures.

- The Gaming Act 1968

This refers to means of escape as one of the requirements for the issue of a licence to operate.

- The Petroleum (Consolidation) Act 1928

This refers to premises licensed for the storage of petroleum and other flammable materials. Guidance should be obtained from the local fire officer regarding the orders and regulations made under this Act.

- The Explosives Acts 1875 and 1923

Acts primarily concerned with licensing and control but also describing precautionary measures; complicated Acts certainly justifying early consultation with the Fire Prevention Officer.

- The Fireworks Act 1951

  Sets out standards for storage in terms of location, construction and quantity of contents.

Appendix 2

LEGISLATION FOR INNER LONDON

- London Building Acts (Amendment) Act 1935

Gives power to modify or waive byelaws including matters concerning fire.

- London Building Acts (Amendment) Act 1939

Part III: Construction of buildings.
Section 16 and 17 – party walls. Section 20 – buildings over 100 ft high or over 80 ft if the building has an area exceeding 10 000 sq. ft or over 250 000 cu. ft if used for trade. (GLC's Code of Practice refers).
Section 21 – openings in party wall.
Section 26 – all public buildings and hotels, hospitals and schools exceeding 250 000 cu. ft – must be to the district surveyor's approval instead of the requirements of the byelaws.

Part IV: – Covers temporary buildings.

Part V: Means of escape in case of fire
Section 34 – new buildings as defined in Section 33, which includes buildings erected after January 1940 and old buildings which have been altered and extended.
Section 35 – the service of notice upon owners of buildings to provide means of escape.
Section 36 – deals with shops projecting beyond the building line.
Section 37 – means of escape to roofs.
Section 38 – buildings used for storage of flammable liquids.
Section 133 – maintenance of means of escape.
Sections 134 and 139 – alterations and change of use.
Sections 145, 150 and 151 – identify exempted buildings.

- London Building (Constructional) Byelaws 1972

  The principal byelaws now in force but Part XI dealing with fire now replaced by the 1979 byelaws (below).

- London Building (Constructional) Amending Byelaws 1979

  Part VI: Fire resistance and combustibility of external walls and claddings, proximity to boundaries and adjoining buildings, size and positions of openings in external walls, fire resistance and combustibility of roofs and coverings.

  Part XI: Fire resistance of elements of construction, separation between buildings. (Similar to but simpler than Part E of the Building Regulations.)

- The London Government Act 1963

  Created the GLC making it the fire and licensing authority for theatres; Petroleum Act, etc.: relates to places of entertainment and exhibition.

- The Greater London Council (General Powers) Act 1966

  Also relates to places of entertainment and exhibition.

- GLC (General Powers) Act 1968

  Relates to the registration of night cafes and their requirements for fire precautions and means of escape and to combustible materials stored in the open.

- GLC (General
- Powers) Act 1975

  Concerns the storage of hazardous materials and provision of warning signs.

- GLC (General Powers) Act 1976

  Provision for take-away food shops open after midnight.

- GLC (General
- Powers) Act 1978

  Amends 1968 Act with regard to flammable materials.

- London Gas Undertaking (Regulations) Acts 1939 and 1954

  Controlling gas installations.

Note: *Fire related legislation for England and Wales will also apply in the Inner London Area.*

Appendix 3

LEGISLATION FOR SCOTLAND

General legislation – Scotland

- Building Standards (Scotland) Regulations 1971 – 1974

  Part D: Structural Fire precautions, fire resistance of structure and materials, protection of ducts, cavities etc and siting of building.

  Part E: Means of escape, smoke control and active measures of fire protection.

  Part F: Chimneys, flues and heat producing installations.

- Fire Precautions Act 1971

  Generally as for England and Wales, except that Section 14 instead of Section 13 applies, and this prevents the imposition of standards higher than those required in the Building Standards. The Designation Orders for various building types apply as in England but with special Orders for hotels and boarding homes.

- The Health and Safety at Work (etc.) Act 1974

  As for England and Wales but amended with respect to the Building Regulations and places certification of hotels, factories, shops and offices with the fire authority.

- Gas Safety Regulations 1972

  Section 22 concerns gas services to buildings.

- Fire Precautions (Special Premises) Regulations 1976

  Concerns safety measures in premises of special fire or explosion risk, e.g. petro-chemical plant.

Local Acts – Scotland

A few local Acts remain for the purposes of control over special premises and licensing of public entertainment buildings. They are administered by the local building control departments using the Building Standards and other relevant Acts as standards.

● The Civic Government (Scotland) Bill

Now before Parliament, this will repeal all local legislation and introduce a single standard of control for the whole of Scotland.

Fire related legislation – Scotland

● The Housing (Scotland) Act 1966.

● Education Scotland Acts 1962 and 1969.

● Social Work (Scotland) Act 1968.

● Nursing Homes Registration (Scotland) Act 1938 and Amendments.

● The Cinematograph (Safety) (Scotland) Regulations 1955.

● The Theatres Act 1968 and Borough Police (Scotland) Act 1892 and Amendment.

● Safety of Sports Ground Act 1975 and Safety of Local Authority Sports Grounds (Scotland) Regulations 1976.

● Gaming Act 1968.

● Licensing (Scotland) Act 1976.

● Petroleum (Consolidation) Act 1928.

Appendix 4

LEGISLATION FOR NORTHERN IRELAND

General legislation – Northern Ireland

The Building Regulations for Northern Ireland (SR 1982 no. 81) varies little from the Building Regulations for England and Wales and any aspects of special uses not covered are dealt with by other legislation which has fire related sections.

The Fire Precautions Act does not apply in Northern Ireland.

Apart from fire related legislation which is in the hands of the local authorities, most fire protection measures are the outcome of voluntary consultation between the local fire authority and the building owners or their designers. The authority also offers documentary guidance in the form of statutory instruments (SIs) relating to Northern Ireland, Codes of Practices and guides published by the Stationery Office for government departments.

Fire related legislation – Northern Ireland

- The Housing (NI) Order 1981 S109.

- Education Act 1944.

- Office and Shop Premises Act (NI) 1966 Sections 28–40.

- The Local Government Act 1934 – (Provisions for Public Entertainments Licensing).

- Fire Services Act (NI) 1969 Sections 10–14A and SI 1975 no. 601.

- The Factories Act (NI) 1965 and SRO NI 1967 nos. 47 and 48.

- The Cinematograph Act 1909 – 59 and Regulations.

- The Petroleum Act 1929.

- The Highly Inflammable Liquids and LPG Regulations (NI) 1974 SRO 256.

Appendix 5

BUILDING RESEARCH ESTABLISHMENT DIGESTS RELATING TO FIRE

APPLICATION FOR
A FIRE CERTIFICATE

APPLICATION FOR A FIRE CERTIFICATE   FP 1(Rev)

Fire Precautions Act 1971

FOR OFFICIAL USE ONLY

---

*To the Chief Executive of the Fire Authority**

Dear Sir

      I hereby apply for a fire certificate in respect of the premises of which details are given below. I make the application as, or on behalf of, the occupier/owner of the premises.

Yours faithfully,

Signature ....

Name ....
             (in block capitals)

If signing on behalf of a company or some other person, state capacity in which signing.

....

Address ....

Telephone Number .... Date ....

To be completed by the Applicant:-

1.   Postal address of the premises

---

2.   Name and address of the owner of the premises: (in the case of premises in plural ownership, the names and addresses of all owners should be given.)

---

3.   Details of the premises
     (If the fire certificate is to cover the use of two or more premises in the same building, details of each premises should be given on a separate sheet.)

    (a) Name of occupier (and any
         trading name if different)

    (b) Use(s) to which premises put

    (c) Floor(s) in building on
         which premises situated

---

*In the case of Crown premises substitute H.M. Inspector of Fire Services

         /over

APPENDIX 6

3. (contd.)

    (d) Maximum number of persons employed or proposed to be employed to work in the premises at any one time

        (i)   below the ground floor of the building   ....

        (ii)   on the ground floor of the building   ....

        (iii)   above the ground floor of the building   ....

        (iv)   in the whole of the premises   ....

    (e) Maximum number of persons other than employees likely to be in the premises at any one time   ....

    (f) Number of persons (including staff, guests and other residents) for whom sleeping accommodation is provided in the premises

        (i)   below the ground floor of the building   ....

        (ii)   above the first floor of the building   ....

        (iii)   in the whole of the premises   ....

---

4. If the premises consist of part only of a building, the uses to which the other parts of the building are put:

---

5. Number of floors in the building in which the premises are situated	6. Approximate date of construction of the premises ....

---

7. Nature and quantity of any explosive or highly flammable materials stored or used in or under the premises

Materials	Maximum quantity stored	Method of storage	Maximum quantity liable to be exposed at any one time

(Continue on a separate sheet if necessary)

112982 500M 3/77 Mcr.(8082)

170

FIRE AND BUILDING

FIRE PRECAUTIONS ACT 1971

NOTES ON COMPLETING APPLICATION FORM FOR A FIRE CERTIFICATE

1.   A fire certificate is required under the Fire Precautions Act 1971 in respect of any premises which are put to a use designated by Order made by the Secretary of State under Section 1 of the Act. Orders have so far been made relating to hotels and boarding houses, factories, offices, shops and railway premises.

Hotels and Boarding Houses

2.   A fire certificate is required for any premises used as a hotel or boarding house if sleeping accommodation is provided there for more than 6 persons (whether guests or staff) or there is some sleeping accommodation above the first floor or below the ground floor.

Factories, Offices, Shops and Railway Premises

3.   A fire certificate is required for any premises used as a factory, office, shop or railway premises in which:—

(a) more than 20 persons are employed to work at any one time; or

(b) more than 10 persons are so employed elsewhere than on the ground floor.

Additionally a fire certificate is required if a building contains 2 or more premises of this kind and the aggregate of persons employed to work in such premises in the building exceeds 20 or the aggregate number so employed elsewhere than on the ground floor in these premises exceeds 10.

4.   A fire certificate is also required in respect of factory premises in or under which explosive or highly flammable materials are used or stored unless the fire authority has determined that the issue of a fire certificate is not justified (see paragraph 9).

Definition of Factory, Office, Shop and Railway Premises

5.   The Designation Order defines these premises by reference to the definitions contained in the Factories Act 1961 and the Offices, Shops and Railway Premises Act 1963, subject to the addition in the one case of electrical stations and institutions and the deletion in the other of premises consisting of a covered market place wherein shop premises are aggregated. If you are uncertain whether your premises come within the definitions you should consult the fire authority for your area.

Application for a Fire Certificate

6.   If your premises need a fire certificate, the application form should be completed and sent to the fire authority for the area in which the premises are situated. If the premises are owned or occupied by the Crown the completed application form should be sent to HM Inspector of Fire Services, Home Office, London SW1 (in Scotland, HM Inspector of Fire Services, Scottish Home and Health Department, St Andrew's House, Edinburgh EH3 3DE).

Who should complete the Application Form

7.   The occupier of premises should complete the application form and send it to the fire authority (or HM Inspector of Fire Services). If, however, the premises are factory, office, shop or railway premises and are held under a lease and consist of part of a building, all parts of which are in the same ownership, or consist of part of a building in which different parts are owned by different people, the Fire Precautions Act 1971, as modified, makes the owner(s) of the building responsible if the premises are used without a fire certificate being in force. Thus in such cases application should be made by or on behalf of the owner(s) of the building. The occupiers of such premises are advised to check that a fire certificate is in force or that an application has been made for one.

APPENDIX 6

Guidance on Completion of the Application Form

8.　(a)　The person completing the application form should state whether the application is made by or on behalf of the occupier/owner, and, if he completes on behalf of a company or some other person, the capacity in which he is signing.

　　(b)　If the premises are in a building which is in plural ownership the names and addresses of all owners should be given.

　　(c)　If the application is for a fire certificate to cover a number of premises in the same building, information as to other premises, as required in Question 3, should be given on a separate sheet.

　　(d)　All the designated uses (see paragraph 1) to which the premises are put should be included in the answer to Question 3(b).

　　(e)　Questions 3(d) and (e) are designed to give information about the occupants of the premises where this is relevant to the scope of a Designation Order.

　　(f)　It is recognised that the occupier of premises may not be able to complete Questions 4 and 6 accurately, but they should be completed to the best of his ability.

Factories in or under which Explosive or Highly Flammable Materials are Stored or Used

9.　Question 7 is designed to provide information about the nature and quantity of any explosive or highly flammable materials stored or used in or under the premises and is particularly relevant in the case of factories. There is a requirement for all factory premises in or under which explosive or highly flammable materials are stored or used to have a fire certificate except in cases where fire authorities determine that the materials are of such a kind and in such a quantity that they do not constitute a serious additional risk to persons in the premises in case of fire. The question on the application form is designed to provide as much information as possible to assist fire authorities in determining whether or not exemptions are justified. It is particularly necessary to give information as to the method of storage and to indicate whether or not the materials are stored in closed containers and also whether or not they are stored in a fire resisting store within the premises. Unless the fire authority has announced any general exemption for small quantities of explosive or highly flammable materials, it will be necessary for occupiers/owners of factory premises, having explosive or highly flammable materials in or under the premises, to apply for a fire certificate. Otherwise they could be in breach of the law.

Further Information and Guidance

10.　If further information is required whether in the case of premises thought to be within the scope of the Designation Orders, an application needs to be made for a fire certificate, any inquiry should be addressed to the appropriate authority specified in paragraph 6. Further guidance on any matter can also be obtained from the authority or from the following guides issued by the Home Office and Scottish Home and Health Department:

No 1	Hotels and Boarding Houses
No 2	Factories
No 3	Offices, Shops and Railway Premises

Plans of premises

11.　On receipt of an application for a fire certificate the fire authority may require plans of the premises to be furnished. These need be no more than a single line drawing of the premises. If however the plans are not provided within the time limit given by the fire authority the application for a fire certificate will be deemed to have been withdrawn.

Inspection of premises

12.　Subsequent to the receipt of plans the fire authority will inspect the premises to assess the adequacy of the fire precautions. If they are not satisfied they will issue a notice specifying the steps that need to be taken before they will issue a fire certificate. If anyone is aggrieved at these requirements he has the right of appeal to a magistrates' court (*in Scotland the Sheriff*) under Section 9 of the Fire Precautions Act 1971.

A FIRE CERTIFICATE

FIRE PRECAUTIONS ACT 1971

FIRE CERTIFICATE

FOR

Identification of Premises
covered by the certificate:

The use or uses of the
premises covered by
this certificate

---

1   IT IS HEREBY CERTIFIED that:

(a)   the premises described above (being premises in respect of which it is
      desired that the above-mentioned uses designated under Section 1 of the
      Fire Precautions Act 1971 be covered by this certificate) are provided
      with the MEANS OF ESCAPE IN CASE OF FIRE specified in this certificate;
      and that

(b)   the relevant building, that is the building containing the premises, is
      provided with the MEANS (other than means for fighting fire) FOR SECURING
      THAT THE MEANS OF ESCAPE with which the premises are provided CAN BE
      SAFELY AND EFFECTIVELY USED AT ALL MATERIAL TIMES specified in this
      certificate; and that

(c)   the relevant building is also provided with the MEANS FOR FIGHTING FIRE
      (whether in the premise or affecting the means of escape) for use in case
      of fire by persons in the building, and with the MEANS FOR GIVING to
      persons in the premises WARNING IN CASE OF FIRE specified in this
      certificate; and that

(d)   the location and quantities of explosive or highly flammable material,
      if any, stored or used in or under the premises are as specified in
      Schedule 2.2.

2   The ALLOCATIONS OF RESPONSIBILITY and REQUIREMENTS in Schedule 2 are
     HEREBY IMPOSED.

3   The Schedules, Standard Terms, Key to Plan Symbols and the Plan(s)
     attached all form part of this certificate.

Date   Authorised
                                        to sign this document

GLOSSARY

Adjuster

Acts for the insurer by financial adjustment of indemnity or reinstatement insurance claims, in accordance with the insurance contract and making recommendations for payments from the insurer to the insured.

After-glow

Process of continued burning with incandescence after flame is extinguished.

Assessment

Financial calculation of the insurance claim.

Assessor

Acts for the insured by preparing an insurance claim based on indemnity rather than reinstatement and negotiating a financial settlement with the adjuster acting for the insurer.

Average

Relates the sum insured to the value of the material asset at risk and, where the former is less than the latter, reduces the amount of compensation in proportion to that shortfall.

Betterment

Financial quantification of the improvement brought about by use of new materials in place of old, or updating of design.

Bund (fire wall)

Enclosure around plant or tanks containing flammable liquid, formed by the erection of walls or banks above ground level safely to contain leakage or spillage.

Burn   To consume or be consumed by rapid oxidation with the production of heat, usually with incandescence or flame, or both.

Charring   Formation of a light, friable, mainly carbonaceous residue of wood or other organic matter, resulting from incomplete combustion.

Chemical foam   Foam formed by the reaction of alkaline salt solution in the presence of a foam stabilizing agent.

Chimney effect   Upward thrust of hot gases by convection currents confined within a vertical enclosure.

Combination meter   Valve assembly which, when hydrants are opened, automatically permits an increased and separately metered supply of water to be drawn from a consumer's branch water main.

Combustible   Capable of burning (opposite is non-combustible).

Combustion   See 'Burn'.

Compartmentation   Division of a building, or large sections of a building, into fire-tight compartments by fire resisting elements of building construction, in order to contain a fire within the compartment of origin.

Compartment floor/wall   Fire-resisting floor/wall used in dividing a building into separate compartments, as defined in the Building Regulations.

Consequential loss   A secondary loss arising from the principal incident (e.g. loss of business profits following material damage to a factory).

Convection (of heat) — transfer of heat in or by a liquid or gas by the movement of the medium.

Damper, fire-resisting — Movable closer within a duct, designed to act automatically to prevent the passage of fire. Together with its frame it must be capable of satisfying, for a stated period of time, the criteria of fire resistance.

Dead end — Area from which escape is possible in one direction only.

Designated use — The purpose for which a building or part of a building is used, or is intended to be used, the purpose groups being identified in the relevant Building Regulations.

Discharge rate — Rate at which a single file of persons can pass through one unit of exit width.

Door, fire resisting — Door which, together with its frame, is capable of satisfying for a stated period of time, the criteria of fire resistance with respect to collapse, flame penetration, and excessive temperature rise.

Drencher head, open or closed — A water discharge assembly fitted to the pipework of a drencher system, open if the pipework is kept in a dry state, closed and controlled by a detection system if the pipework is kept charged with water.

Drencher system — A system of water pipes with drencher heads (open or closed, see above) fitted at suitable intervals and heights, designed so as to protect large surfaces from fire exposure by the discharge of water (usually in heavy quantities) and generally controlled by a fire detection system

Dry powder

Extinguishing medium consisting of finely ground chemicals comprising a main constituent, such as sodium bicarbonate, potassium bicarbonate or ammonium phosphate, and additives to improve its flow, storage and water repellent characteristics.

Dry riser

Vertical pipe installed in a building for fire fighting purposes, fitted with inlet connections at fire brigade access level and landing valves at specified points. It is normally dry but capable of being charged with water, usually by pumping from fire service appliances.

Eighty five percent rule

Additional clause attached to, or inserted in, an insurance policy, excluding application of average, provided the sum insured is at least 85% of the total reinstatement cost of the building. (Usually applies only in a situation of partial loss of less than 85%.)

Escape route

A route forming part of the means of escape from a point in a building to a final exit.

Evacuation procedure

Predetermined plan of action designed to achieve the safe evacuation of the occupants of a building.

Evacuation time

Time taken for all occupants of a building or part of a building, on receipt of an evacuation signal, to reach final exit.

Excess

Financially the first part of any risk excluded from the policy: a sum stated in the policy which is deductible from any claim.

Explosive range	Range of percentage of vapour (gas) concentration by volume in air between the upper and lower limits of flammability.
Exposure hazard	Risk of fire spreading from a building to an adjacent separate building, or to another part of the same building, by radiated heat across the intervening space.
Extinguish (fire)	To bring a fire to an end by one or more of the following: starvation, inhibition, smothering or cooling, achieved by the reduction of fuel, oxygen and temperature.
Extinguishing system	A layout of pipes to distribute a fire fighting agent throughout a building, fitted with nozzles at suitable intervals and heights through which the agent is discharged. May be used with any of the following agents: carbon dioxide (CO_2) dry powder foam inert gas.
Fire	Process of combustion characterised by heat or smoke or flame, or any combination of these.
Fire alarm system	System of fixed apparatus for giving an audible and/or visible and/or other perceptible warning of fire and which may also initiate other action.
Fire authority	The body authorised to act in the enforcement of fire regulations and legislation.
Fire blanket	Blanket specifically designed to be used for smothering small fires.
Firebreak	Open space separating buildings, stored products or other combustible materials, which is

capable of restricting the spread of fire.

Firebreak wall/floor F.O.C. term to define a fire resisting wall/floor.

Fire certificate Certificate issued by the fire authority certifying that the fire regulations have been complied with for a specific use and stating requirements as to maintenance of fire safety standards.

Fire classification System of classifying fires in terms of the nature of the fuel. *Note* The following terms and definitions are taken from BS 4547 (EN 2):

Class A fire – Fires involving solid materials, usually of an organic nature, in which combustion normally takes place with the formation of glowing embers.
Class B fire – Fires involving liquids or liquefiable gases.
Class C fire – Fires involving gases.
Class D fire – Fires involving metals.

Fire detector Device which gives a signal in response to a change in the ambient condition due to fire. May be of the following types:

gas fire detector Detector which responds to the presence of gases produced by combustion or thermal decomposition.

fixed temperature detector Detector which responds to an
heat detector increase in temperature.
heat fire detector
rate of rise detector

infra-red detector Radiation fire detector which responds to infra-red radiation.

ionization chamber smoke fire detector
Smoke detector which responds when smoke, having entered the detector, causes a change in ionization currents within the detector.

laser fire detector
laser detector
Detector which responds to the effect of fire upon the laser beam.

line fire detector
Detector which responds to fire detected anywhere along its length.

optical smoke fire detector
Detector having a photo-electric cell which responds when light is absorbed or scattered by smoke particles.

radiation fire detector
Detector which responds to radiation emitted by a fire.

Fire exit
Way out of a room or space to be used by the occupants in the event of fire; also terminal point of an escape route beyond which persons are no longer in danger from fire.

Fire extinguisher
Portable item of fire fighting equipment containing a fire extinguishing medium which is expelled by internal pressure or by the action of a self-contained hand pump.

Fire fighting access
Approach facilities provided to and/or within a building to enable fire service personnel and equipment to gain access.

Fire lift
Lift installed for normal purposes in a building but provided with a priority switch for fire brigade use.

Fire load
Total amount of combustion material expressed in heat units, or its equivalent weight of wood.

Fire load density
Fire load per unit area.

Fire Offices Committee (F.O.C.)　　Committee of representatives from fire insurers (see also F.O.C. Rules)

Fire path　　Access route from highway for fire fighting or rescue appliances.

Fire point　　Location where fire fighting equipment is sited which may also comprise a fire alarm call point and fire instruction notices, the whole being provided and arranged for the use of occupants of premises.

Fire propagation index　　Comparative measure of the contribution to the growth of fire of a combustible material, as determined in accordance with BS 476, Part 6.

Fire resistance　　Ability of an element of building construction to satisfy for a stated period of time some or all of the criteria specified in BS 476, Part 8, namely resistance to collapse, resistance to flame penetration and resistance to excessive temperature rise on the unexposed face.

Fire resisting　　Having the ability to withstand the effects of fire for a specified period of time without loss of its fire separating or loadbearing function, or both.

Fire resisting separation　　Separation of parts of a building or structure by means of fire resisting walls and floors or similar related elements of structure.

Fire retardant　　Substance or treatment applied to a material to increase its resistance to destruction by fire. (Also used as adjective.)

Fire valve　　Valve provided on a pipe carrying flammable liquid or gas, designed

	to cut off the supply automatically or by manual over-ride on the occurrence of a fire.
Fire vent	Automatic and/or manually operated vent in the enclosing walls or roof of a building for releasing heat and smoke in the event of fire.
Fire wall	Wall, screen or separating partition erected in the open air to reduce or avoid risk of radiated heat from or to a building, structure, plant, piece of apparatus or stock pile.
Fireman's switch	Switch provided for the use of firemen by means of which high voltage equipment can be isolated from the supply during fire fighting.
Flame	Zone of oxidation of gas usually characterised by the liberation of heat and the emission of light.
Flame retardant	Substance or treatment applied to a combustible material to decrease its tendency to propagate flame across its surface. (Also used as adjective.)
Flammable (preferred to inflammable)	Capacity to burn with a flame (opposite is non-flammable).
Flammable gas detector	Device that monitors the atmosphere in a specific area and indicates the concentration of flammable gas present.
Flammability range	See 'Explosive range'.
Flash back	Rapid movement of flame through a flammable vapour/air mixture from the point where it is ignited to the point where it is generated.
Flash-over	Stage in the development of a con-

tained fire at which fire spreads rapidly to give large merged flames throughout the space. As a scientific term 'flash-over' is applicable only to enclosed compartments.

Flash point

Lowest temperature at which a liquid gives off sufficient flammable vapour in air to produce a flash on the application of a small flame.

Foam compound

Mixture of foam forming materials, such as protein or fluorinated surface active agents, which produce mechanical foam used to smother fires or suppress the emission of gases.

Foam inlet

Fixed equipment consisting of an inlet connection, fixed piping and a discharge assembly, enabling firemen to introduce foam into an enclosed compartment.

Foam pourer

Discharge assembly, either transportable or fixed to the rim of a tank of flammable liquid, which enables foam to be introduced into the tank to form a layer on the surface of the liquid.

F.O.C. Rules

Rules compiled by the Fire Offices' Committee for defining grades of construction, used by insurers to assist in determining the degree of risk and the premium rate of a building.

Fusible link

Soldered connecting link designed to part on melting at a specified temperature.

Hydraulic platform

Fire appliance consisting of a platform or cage, supported on hydraulically powered manoeuvrable arms, from which water or foam can be directed on to a fire or with which persons can be rescued.

Ignitability

Ability of a material to be ignited by a small flame, as determined in accordance with BS 476, Part 5.

Ignition temperature

Lowest temperature of a substance at which sustained combustion can be initiated.

Incandescence

Emission of light by a substance due to its high temperature

Indemnity cash payment

Payment made in settlement of a claim based purely on a financial assessment of the loss and intended to put the insured back, after the loss, into a similar position to that before the incident (the assessment after adjustment by the loss adjuster).

Indemnity value

Value of the material asset used as the sum insured – often assessed by depreciating total reinstatement costs by deduction for wear and tear.

Inerting Agent

A powder or gas (such as nitrogen or CO_2) which suppresses fire by the physical means of displacing the air in the area of combustion so as to cut off the oxygen supply, or by reducing the temperature below the combustion level.

Inflammable

See 'Flammable'.

Inhibiting Agent

Usually a gas which combines inertion with chemical suppression of the combustion process.

Inhibition

Process of extinguishing fire by use of an agent which interrupts the chemical reactions in the flame.

Insurance broker

Advises on all aspects of insurance

and places the insured risks with insurers on the most favourable terms.

Insurance surveyor
Acts for the insurers in evaluating the risk on a material asset, often assisted by application of the F.O.C. Rules.

Insured
Pays the premium to the insurer and is relieved of the risk.

Insurer
Receives the premium from the insured and takes the risk.

Latent heat
Heat required to change the state of a substance without causing a change in temperature.

Local authority clause
Additional clause attached to, or inserted in, an insurance policy providing for the additional cost in reinstatement of complying with current Building Regulations and other local authority require-ments, including planning.
(*Note* prevents the operation of betterment but the sum insured must be adequate to cover the additional work.)

Loss adjuster
See 'Adjuster'.

Lower limit of flammability
Lowest percentage concentration, by volume, of flammable vapour (gas) mixed with air which will burn with a flame.

Material asset
The building or item at risk in the insurance policy.

Means of escape
Structural means whereby a safe route is provided for persons to travel from any point in a building to a place of safety by their own unaided efforts.

Non-combustible
(preferred to incombustible)
Not capable of burning.

Non-flammable (preferred to inflammable)	Not capable of burning with a flame.
Obsolete buildings clause	Additional clause attached to, or inserted in, an insurance policy providing a declared reduction in the sum insured so that the sum represents the value of an appropriate replacement building rather than a replica of the existing obsolete building. Prevents the application of average in this respect.
Oxidation	Combination of oxygen with a substance.
Pressurised escape route	Escape route within which the air is maintained at a higher pressure than that in the remainder of the building.
Protected area	Area giving an adequate degree of protection from fire in another area and from which there is means of escape.
Protected corridor	Corridor forming part or whole of the horizontal component of a protected escape route.
Protected lobby	Lobby having an adequate degree of protection and forming part or whole of the horizontal component of an escape route, or affording additional protection to an escape route.
Protected premises	Premises or part of premises provided with an automatic system or systems for detecting and/or extinguishing fire and usually having means of transmitting a fire signal to a remote manned location.
Radiation (of heat)	Transfer of heat through a gas or vacuum other than by heating of the intervening space.

Reconstruction	Rebuilding after damage incorporating variations from the original design or construction.
Reinstatement	Re-establishing the material asset to its former position and restoring it in proper order – 'new for old'.
Reinstatement memorandum	Additional clause attached to or inserted in an insurance policy providing for the cost of reinstatement to take place without application of betterment – 'new for old'.
Roof, fire resisting	Roof construction which, when subjected to conditions of internal fire, is capable of satisfying for a stated period of time the criteria of fire resistance with respect to collapse and flame penetration.
Roof, resistance to external exposure	Ability of a roof deck and covering to resist both penetration by external fire and flame spread over the external surfaces, as determined in accordance with BS 476, Part 3.
Roof screen	Vertical screen fitted internally in the roof of a building to divide the roof into bays, so that smoke and hot gases from a fire are contained within the bay of origin.
Safety curtain	Vertical fire resisting screen, operated by gravity with a quick-release device, designed to close the proscenium opening in the event of fire in a theatre or similar place of public assembly.
Salvage	Procedures to reduce incidental losses from water, smoke, weather and theft during and following fires.

Salvage equipment

Equipment other than fire fighting equipment used to mitigate incidental damage during and following fires.

Schedule of loss

Financial assessment representing the notional rein-statement of damage.

Self-ignition temperature

Temperature at which a flam-mable gas/air mixture will ignite without an external source of ignition.

Smoke

Visible airborne cloud of fine particles, the product of incom-plete combustion.

Smoke outlet

Opening, or fire resisting shaft or duct provided in a building to act as an outlet for smoke and hot gases produced by an outbreak of fire.

Smouldering

Process of combustion without flame but usually with incan-descence.

Specific heat

Amount of heat required to raise temperature of unit mass by 1 degree celcius.

Spontaneous combustion

Biological or chemical reaction which produces its own heat resulting in combustion.

Sprinkler head, open or closed

As drencher head, a water dis-charge assembly fitted to the pipework of a sprinkler system which scatters water automatically in a predetermined pattern over a specified area, open if the pipe-work is kept in a dry state, closed and automatically opened if the pipework is charged with water.

Sprinkler system

A system of pipes with sprinkler heads fitted at suitable intervals and heights, designed so as to control or extinguish fire by the discharge of water in which each of the heads or groups of heads opens automatically at a specified temperature or by the operation of a fire detection system.

Sprinkler water accelerator Device fitted to an automatic dry pipe or alternative sprinkler system for accelerating the flow of water into the installation by causing the control valve to operate at a lower pressure differential than the normal setting.

Stairway, enclosed Stairway in a building, physically separated (by walls, partitions, screens, etc.) from the accommodation through which it passes, but not necessarily a protected stairway.

Stairway, protected Stairway having an adequate degree of protection and forming the vertical component of a protected escape route.

Starvation Process of extinguishing fire by limitation or reduction of fuel.

Sum insured The maximum sum at risk in the insurance policy.

Surface spread of flame classification Division into classes of combustible building materials according to the rate at which flame spreads over their surfaces, as determined in accordance with BS 476, Part 7.

Thermocouple Junction of wires of dissimilar metals used for measuring temperature.

Thermostat Automatic control device responsive to changes of temperature.

Turntable ladder Fire appliance embodying a manoeuvrable extending ladder from which water or foam can be directed on to a fire or with which persons can be rescued.

Valuation for fire insurance purposes

Financial assessment of the value of a material asset used as the basis for ascertaining the sum insured.

Ventilated lobby

Protected lobby provided with means of ventilation to the open air.

Volatility

Readiness with which a substance vaporises.

Wall, fire resisting

Wall, either loadbearing or non-loadbearing, capable of satisfying, for a stated period of time, the criteria of fire resistance with respect to collapse, flame penetration and excessive temperature rise.

Wet riser

Vertical pipe installed in a building for fire fighting purposes and permanently charged with water from a pressurised supply, fitted with landing valves at specified points.

Wheeled escape

Wheeled extending ladder, usually mounted on a fire appliance from which it can be removed and manoeuvred into position for rescue or fire fighting purposes.

INDEX